The Evolutionists

The Struggle for Darwin's Soul

The Evolutionists

*The Struggle for
Darwin's Soul*

Richard Morris

W. H. Freeman and Company
New York

Text Design by Blake Logan

Library of Congress Cataloging-in-Publication Data
Morris, Richard.
   The evolutionists: the struggle for Darwin's soul / Richard Morris.
      p.   cm.
   Includes bibliographical references (p.   ).
      ISBN 0-7167-4094-X (hardcover)
      1. Evolution (Biology) 2. Natural selection. I. Title
      QH366.2 .M683 2001
      576.8'2—dc21

                                                          00-011832

Printed in the United States of America

First printing 2001

W. H. Freeman and Company
41 Madison Avenue, New York, NY 10010
Houndmills, Basingstoke RG21 6XS, England

contents

Preface   vii

1 Controversy in Evolutionary Biology   1

2 The Fossil Record and the Evidence for Evolution   13

3 Darwin's Five Theories of Evolution   47

4 Darwinian "Fundamentalism"   75

5 How Gradual Is Evolution?   99

6 The Sciences of Complexity   125

7 Evolutionary Psychology   157

8 The Evidence   195

9 Controversy and Discovery   225

A Selected, Annotated Bibliography   239
World Wide Web Resources   245
Index   252

I f you are not a scientist, the chances that you have ever gone to a scientific conference are small indeed. If you were to attend one, the proceedings would probably not be quite what you expected. Oh, you would see scientists giving papers in which they report their latest results. But you would also hear a lot of arguments. When a new finding is announced at a conference, members of the audience often raise objections or question the validity of the work being described.

In other words, you would witness a certain amount of controversy. Perhaps you would be surprised by this. But none of the scientists in attendance would be. They all know that scientific controversy is as old as science itself. Scientists are a skeptical lot. Contrary to what many people think, most of their skepticism is not directed toward such things as astrology or the predictions of psychics. Most scientists expend much more energy expressing skepticism about claims made by other scientists.

This is as it should be. When a new scientific theory is propounded or when experimental results that throw doubt on an old theory are presented, scientists look for holes in the arguments or flaws in the experimental design. They accept nothing until they conclude that the evidence is convincing. They must do this if science is to progress. Scientific knowledge cannot be built on dubious ideas and flawed experimental results.

All the major scientists in history have become embroiled in controversies. Some, like Galileo, reveled in it. Others have avoided it. But all
have found themselves involved in disputes of one kind or another—not
always disputes that they seek or instigate. For example, Isaac Newton went
so far as to withhold his findings to avoid becoming involved in controversies. His great work, *Philosophiae naturalis principia mathematica*
(*Mathematical Principles of Natural Philosophy*), which contained his
laws of motion and his law of gravitation, practically had to be pried away
from him. Holding on did him little good: controversies arose nevertheless.

You probably won't be surprised if I tell you that there was a lot of
controversy after the publication of Charles Darwin's *On the Origin of
Species* in 1859. You may not be aware, however, that much of the debate
did not center around the issue of whether or not evolution had indeed
taken place. It is true that large segments of the general public remained
unconvinced. Even today, about 50 percent of the population of the United
States does not believe in evolution. However, by the mid-1860s, most
British biologists had accepted the truth of evolution. Many of the debates
that took place then were between scientists who accepted evolution but
who felt that there was reason to argue about the details of the theory.

Some of Darwin's staunchest supporters disagreed with him on key
points. For example, the British biologist T. H. Huxley earned the name
"Darwin's Bulldog" for his spirited and unrelenting efforts to convince
scientists and members of the lay public alike of the truth of evolution.
Huxley wrote one article after another about evolution and sent them to
the leading periodicals of the day. He defended Darwin against criticisms,
and replied to unfavorable reviews of Darwin's book. Yet Huxley's views
differed from Darwin's in several different respects.

For example, Darwin attributed evolution to the action of natural
selection. Huxley wasn't so sure about this. He believed that other factors

might play a role. He spoke of evolution in "predetermined" directions, an idea that appeared nowhere in Darwin's writings. And he expressed the opinion that evolution was not always so gradual a process as Darwin conceived it to be.

Huxley was not the only evolutionist who disagreed with Darwin. The botanist Joseph Hooker, another strong supporter of evolutionary ideas, also took issue with Darwin about the specifics of the theory. So did the naturalist Alfred Russel Wallace, who had discovered the idea of natural selection independently. All these men argued with Darwin privately in letters and sometimes in print.

Most of the disputes between Darwin and other British scientists are nearly forgotten today. The issues that were raised have, for the most part, been settled. Many of these issues were raised because neither Darwin nor his contemporaries knew anything of genetics. As a result, there were unanswered questions about natural selection. Some pioneering work in genetics had been done by the Austrian monk Gregor Mendel. However, Mendel's publications on the subject never came to the attention of leading evolutionary biologists, and decades were to pass before they were rediscovered.

A great deal of scientific work was done on genetics during the first half of the twentieth century. Genetics provided scientists with an understanding not only of the mechanism of biological inheritance but also of the role played by mutations. Once this knowledge became widespread within the scientific community, the majority of the problems that puzzled Darwin and his contemporaries were solved.

However, controversy about the details of the evolutionary process never stopped. When one question was answered, scientists would inevitably begin arguing about another. And of course this was exactly as it should have been. A scientific discipline in which there is no controversy is one that is dead. As long as scientists keep discovering new knowledge, there will inevitably be controversy about the interpretation of the evidence. Scientific controversies are always settled sooner or later, if they are not simply forgotten. Often, as soon as they settle one argument, scientists embark on another. This is one of the mechanisms of scientific progress.

Today, scientists are arguing about evolution as fiercely as ever. Naturally they are not debating the question of whether or not evolution happened. That has not been an issue since the nineteenth century. Nor are they arguing about the importance of natural selection; all agree that natural selection is the main driving force behind evolutionary change. However, members of opposing camps have been arguing with a ferocity that has rarely been equaled in the annals of science.

Although there is broad agreement among evolutionary scientists on many points, there are important scientific issues that remain unsettled. I have written this book to explain what the arguments are about, to summarize those on each side, to take a look at the relevant empirical evidence, and to suggest what kinds of evidence are needed to bring the controversies to an end.

It would be foolish to come down in favor of one side or another while controversy is still raging within the community of evolutionary biologists, so I have taken no sides in the arguments.

Controversy in Evolutionary Biology

According to the noted British geneticist John Maynard Smith, Harvard paleontologist Stephen Jay Gould is "a man whose ideas are so confused as to be hardly worth bothering with." Oxford University zoologist Richard Dawkins, author of the best-selling book *The Selfish Gene,* charges that Gould's view of evolution is based on a fundamental misunderstanding. Tufts University philosopher Daniel Dennett goes further. According to Dennett, Gould is "a would-be revolutionary" who has mounted a series of attacks on conventional Darwinism over the years. Furthermore, Dennett says, as the best-known writer on evolutionary topics, Gould has had an influence that is "immense and distorting." Gould must have some "hidden agenda," Dennett speculates. Perhaps it is Gould's Marxist leanings, he says, that have caused him to attack accepted evolutionary theory.

Gould, on the other hand, brands Maynard Smith, Dawkins, and Dennett as "Darwinian fundamentalists," who place an emphasis on one component of Charles Darwin's thinking and "push their line with an almost theological fervor." Maynard Smith, he says, has apparently gotten caught up in an "apocalyptic ultra-Darwinian fervor." Dennett's writings, he adds, are characterized by "hint, innuendo, false attribution and error." If the Victorian era British biologist T. E. Huxley had been "Darwin's bulldog," Gould concludes, then perhaps Dennett should be characterized as "Dawkins's lapdog."

Maynard Smith, Dawkins, Dennett, and Gould are not the only individuals engaged in this controversy. For example, Gould's colleague, paleontologist Niles Eldredge of the American Museum of Natural History, has also criticized Dawkins, Dennett, and Maynard Smith. So have various other scientists, such as geneticist H. Allen Orr of the University of Rochester in New York. Evolutionary psychologists such as Steven Pinker of the Massachusetts Institute of Technology and Leda Cosmides and John Tooby of the University of California at Santa Barbara are members of the other camp. The controversy is more than an argument between Gould and his critics. It is one in which numerous scientists are currently engaged because they believe that there are still questions to be settled about the nature of evolution.

Before I go any further, I should point out that none of the participants in this controversy is questioning the idea of evolution. They are all well-known evolutionary biologists with the exception of Dennett, a philosopher who is the author of *Darwin's Dangerous Idea,* which was a finalist for the 1995 National Book Award. They all agree that the evidence that evolution has taken place is overwhelming, and they all oppose creationism, which they consider to be an attack on science. Furthermore, they all agree that Darwin's idea that natural selection is the main mechanism of evolutionary change is correct. And, of course, they all tend to view themselves as Darwin's intellectual heirs.

What are they arguing about, then? And why have the arguments become so bitter? Those are the questions I try to answer in this book. I think I will be able to show you that there is indeed a lot to argue about and that some of the issues that have been raised are important ones.

Darwin's theory of evolution is universally accepted among biologists. However, Darwin's "theory" is not a single idea; some scientists have broken it down into five or more subtheories. Thus it is possible for scientists to agree on many of the details of the theory while arguing about others. Furthermore, it is possible to agree that natural selection is the main cause of evolution, while debating the details of how evolution has happened.

Controversies About the Role of Natural Selection

The idea of evolution was not original with Charles Darwin. On the contrary, it was an idea that had been widely discussed for decades when Darwin published his book *On the Origin of Species by Means of Natural Selection, or the Preservation of Favoured Races in the Struggle for Life* in 1859. Darwin's great achievement was to amass evidence sufficient to convince his scientific contemporaries that evolution had indeed taken place, and to suggest a mechanism, natural selection, that would cause species to evolve.

Darwin realized that the members of every species were capable of producing more offspring than could survive. As Darwin noted, this is true even of such slow-breeding species as the elephant. If all the offspring of a herd of elephants lived to maturity and produced offspring of their own, the world would be overrun with elephants in no more than a few thousand years. Darwin reasoned that some members of a species were more likely to survive and reproduce than others. Those that did would pass their characteristics along to succeeding generations. If there was originally a certain amount of variation in a species, some characteristics would be preserved, while others would be weeded out. Over long periods of time the species would change; gradual evolution would take place.

Natural selection does not require that the better-adapted members of a species *always* survive and reproduce. If they possess characteristics that make survival 1 percent or even a fraction of a percent more likely, then on the average they will have more offspring than other members of the species. Thus if some of the evolutionary ancestors of the giraffe were able to obtain more food than other giraffes because they had longer necks, then long-necked animals would eventually come to predominate and eventually evolve into the giraffe that exists today. If an antelope could run a little faster than others, it would be more likely to escape predators and reproduce. Its offspring most likely would also be a little faster. Eventually that characteristic would spread throughout the population and in time all antelopes would be swifter.

Natural selection is an idea that is beguiling in its simplicity. Once you understand it, it is easy to become convinced that evolution could happen no other way. However, since the time of Darwin, some scientists have thought that natural selection was not the only driving force behind evolution. In Darwin's day, some biologists thought that there were other important factors. Some of them believed, for example, that some inner force drove human evolution toward greater intelligence. This idea was somewhat metaphysical and today it is thought to be totally discredited. However, scientists have never stopped looking for factors other than natural selection that might influence evolutionary change. I'll be discussing a number of such possibilities in the chapters that follow.

Stephen Jay Gould and his colleague Niles Eldredge are perhaps the best-known scientists who proclaim such a belief. Consequently their names—especially that of Gould, who is the more vocal and contentious of the two—are frequently mentioned by those engaged in the controversies that such ideas have caused. Gould and Eldredge believe that natural selection acting on individual organisms is the main cause of evolutionary change. But they also think that there are phenomena that appear at higher levels of complexity. For example, they have proposed a theory of "species sorting." They claim that, just as some individuals are more likely to survive than others, species also have different probabilities of survival. Furthermore, Gould says, some evolutionary lines give rise to more new species than others. Surely, he suggests, this is a factor in evolution too.

Gould's and Eldredge's theory has been attacked by such orthodox Darwinists as Richard Dawkins and Maynard Smith, who continue to insist that only natural selection is important. In reply, Gould has contrasted Dawkins's and Maynard Smith's "Darwinian fundamentalism" with his and Eldredge's "evolutionary pluralism." The controversy is still going on, and it is not likely that the argument will end soon. As I will show later on, the differences between the two groups of scientists do not arise from their interpretations of scientific data. There are fundamental philosophical differences between them. Dawkins, Maynard Smith, and other orthodox Darwinians are reductionists* who see only one important factor in evolution. Gould and Eldredge, on the other hand, describe themselves as pluralists who see evolution as something that is much more complex.

The differences in outlook have led to not one but a variety of different controversies. The first one erupted in 1979 when Gould and Harvard geneticist Richard Lewontin published a paper in which they criticized methods that, they claimed, were commonly used by evolutionary biologists. Rather than looking at organisms as integrated wholes, Gould and Lewontin said, these scientists reductively broke them down into collections of individual "traits" and then invented stories that purported to explain why these traits had evolved. But, the two authors said, there was

---

*This isn't a criticism. As I will explain later, reductionism is a valid scientific method.

often no evidence to support these sometimes fanciful stories. Gould and Lewontin then went on to maintain that organisms often had traits (which they called spandrels, after the architectural term) that had evolved for no adaptive purpose—in other words, traits that had appeared as by-products of other evolutionary changes.

Naturally, the orthodox Darwinists objected. This was all very well understood, they replied. Gould and Lewontin were setting up a straw man and then knocking him down. In any case, it couldn't be denied that even though spandrels didn't have any functions originally, natural selection could then modify them for adaptive purposes. The point that Gould and Lewontin had made was, in their view, one that wasn't terribly important.

Evolutionary Psychology

Eventually the controversy died down. However, the idea of spandrels came to the fore again when Gould began to voice his skepticism about the methods of a new field called evolutionary psychology. Evolutionary psychologists are scientists who attempt to explain human behavior in evolutionary terms. They postulate that the human brain is made up of a collection of "mental modules" that were created by natural selection to cause us to behave in certain specific ways. The evolutionary psychologists claimed to have found modules that produced characteristic human mating strategies, modules that were responsible for cooperative behavior, and so on. Two evolutionary psychologists, Randy Thornhill and Craig T. Palmer, went so far as to maintain that there was evidence that natural selection had produced mental modules that explained male propensity to rape.

The hypothesis advanced by the evolutionary psychologists is not based on any hard neurological evidence. Neurobiology has not advanced to the point where it would be possible to find the neural circuits that correspond to such modules, if indeed mental modules do exist. Thus there is room for controversy. And indeed there has been controversy. Gould in particular has expressed skepticism about the reductionist methods of evolutionary psychology, just as he has objected to what he views as excessive reductionism in the field of evolutionary biology. In his view, human mental functioning cannot be broken down into a collection of distinct traits. He sees the human brain as a kind of general-purpose computer full of spandrels. There was obviously something about the life of human ancestors on the African savannas many hundreds of thousands of years ago, he says, that made big brains advantageous. But once big brains evolved, Gould goes on, they became capable of doing numerous things that were not related to the reasons why natural selection created them in the first place. He points out, for example, that our ability to learn to read and write must be a spandrel. It could not have evolved for some adaptive purpose; when the brain first became large, written language did not exist.

Naturally the evolutionary psychologists did not take kindly to Gould's criticisms. Their replies to Gould have, if anything, been even more bitter than those of the orthodox Darwinists. Some of them have claimed that Gould has misrepresented evolutionary psychology, others have gone so far as to accuse Gould of intellectual dishonesty. This controversy continues to rage with no end in sight.

The Sciences of Complexity

Some of the ideas proposed by Gould and his colleagues are reminiscent of those of the scientists who work in a relatively new field called the sciences of complexity. Complexity scientists come from many different disciplines, including mathematics, physics, computer science, and biology.

They work in the field because they have become interested in exploring what are called the emergent properties of complex systems. Complexity scientists maintain that, although reductionist methods are an essential part of science, there are some problems they cannot solve. Complex systems, they say, cannot be reduced to the properties of their components.

Ant colonies provide a good example of this idea. Ants are creatures with simple nervous systems. An individual ant does not have a complex repertoire of behavior, and ants signal one another in only about a half dozen ways. But an ant colony is capable of very complex kinds of activity. Some species of ants, for example, make slaves. Others maintain fungus "farms." Yet others "milk" aphids that they keep as a kind of domestic animal. But these kinds of behavior come about when large numbers of ants interact. The activities of ant colonies, in other words, are emergent properties that are not present in the individuals of which the colonies are composed.

Complexity scientists typically study the systems in which they are interested by modeling them on computers. Computers models of colonies of ants have been created, for example. The behavior exhibited by the "ants" in a computer can be eerily like the behavior of the real insects. Complexity scientists have also created computer models of flocks of birds, hives of bees, and even ecological systems in which one species preys on another. Some of these models have been quite successful. For example, the flocking behavior of computer birds looked so realistic that the program used to create them aroused interest among ornithologists.

Evolving species can also be viewed as complex systems, and some complexity scientists have begun to create computer models that might tell us something about the ways in which evolutionary change takes place. For example, evolutionary biologist Stuart Kauffman has created models of networks of genes, and biologist Thomas Ray has created artificial digital organisms that are born, reproduce, mutate, and evolve inside computers.

Some of the work done in the sciences of complexity is as controversial as that done by the evolutionary psychologists. Biologists, in particular, have been skeptical of the idea that computer models can really tell us anything of importance about evolution. Not surprisingly, some of the biologists who have criticized complexity science have been the same ones who have been so vehemently opposed to Gould's and Eldredge's pluralist view of evolution. They have expressed the view that it is possible to get anything out of a computer simulation, depending on what is put into it, and they accuse the complexity scientists of paying little attention to empirical facts. Naturally the complexity scientists disagree.

Looking at the Evidence

The recent controversies in evolutionary biology are far from being settled. On the contrary, they are multiplying, and reasoned arguments have increasingly given way to vituperation. One of the reasons is that there has not been sufficient evidence to answer the questions that have been raised. However, evolutionary biology is anything but a stagnant field. New discoveries are made almost every day. Some of them are beginning to shed some light on the issues that are being argued about. At the same time new questions are being raised.

For example, recent field studies have shown that, in some cases, natural selection is more powerful than even advocates of orthodox Darwinism had believed. Scientists have found that, in some species, observable evolutionary change can take place in as little as ten years. Scientists studying North American fruit flies have recently discovered that flies that migrated to different parts of the hemisphere evolved differences in wing length in only a decade. Furthermore, the differences in wing length matched those of flies that lived under corresponding climatic conditions in Europe.

It has also been discovered that speciation—the creation of new species—does not always happen in the manner that most evolutionary biologists thought it did. It had previously been thought that new species evolved when part of a population became geographically isolated. Since the two subpopulations could no longer interbreed, they would, it was assumed, evolve in different directions. However, biologists conducting field studies have recently discovered that populations that are not separated by geographical barriers often split into two different species by adapting themselves to different environments. This has been observed, for example, in stickleback fish living in Canadian lakes. In a number of different locations, similar but distinct species evolved when some of the fish continued to swim in open water, while others became bottom dwellers.

Finally, contrary to what had previously been thought, it appears that evolution does not invariably take place through the gradual accumulation of numerous small, favorable mutations. It has been shown that sometimes new species are created when a small number of genetic mutations have large effects. This has been seen, for example, in fruit flies on an archipelago in the Indian Ocean and in monkey flowers growing in California.

Some of these discoveries are so recent that it is not clear what their implications will be. But if I had to make a guess, I would say that ten years from now evolutionary biologists will still be arguing, most likely about different things. New evidence about evolution is accumulating. Some of it is so surprising that it is likely that new theories will have to be developed to explain it. And theories are something to argue about.

I expect that, in the years ahead, evolutionary biology will become even more of a contentious field than it is now. But of course there is nothing wrong with that. Scientific controversy, though it may be bitter at times, is a healthy thing. It is a sign that scientists are questioning old ideas and looking for ways to better understand existing evidence. Scientific controversy is a sign of vitality. The absence of scientific controversy in a scientific field is a sure sign that few further advances will be made. For example, no one argues about planetary motion anymore, or about how a lens focuses light. The significant questions about those matters were settled long ago, and there is nothing more to add.

Science, in other words, is a quest for knowledge. And when new discoveries are made, it is only natural that scientists should argue with one another about what they have found.

I will be discussing these and other controversies in the chapters that follow. Then I will talk about some of the new evidence concerning evolution that is appearing. But before I do either, I think it would be a good idea to discuss Darwin's theory and the evidence for it in some detail. I will talk about the various subtheories that make up "Darwin's theory of evolution" and try to give you some of the ideas that seem firmly established and that seem to allow for different interpretations. After all, to follow an argument, it is necessary to understand what is being argued about. Fortunately the relevant ideas are simple ones, and it won't be necessary to devote a great deal of space to explaining them.

I will begin discussing the current controversies in evolutionary biology in Chapter 4. Since there are quite a number of them, discussions of them will constitute the bulk of this book. Finally I will talk about some of the new discoveries that have recently been made.

The Fossil Record and the Evidence for Evolution

M any of us have seen fossils of dinosaurs and other extinct animals in natural history museums. As a result, when we think of fossils, our minds create an image of a record of terrestrial vertebrate life. But in reality the fossil record is much more extensive than that. There are more fossils of marine animals than there are of land-dwelling creatures, and there exist fossils of such organisms as plants and microorganisms as well. The dinosaurs died out relatively recently, about 65 million years ago. But 65 million years is less than one-fiftieth of the time that has elapsed since the first life evolved.

The fossil record of the evolution of life on Earth stretches back 3.5 billion years. Fossils of bacteria resembling blue-green algae have been found in rocks of that age in Australia and Africa. Furthermore, there are indications that life may have existed at an even earlier time. Chemical traces of life—substances that look as though they were formed by living organisms—have been found in rocks in Greenland that have an age of 3.85 billion years. Since it is thought that the surface of Earth remained molten until approximately 3.8 billion years ago or a little earlier, there are indications that life existed on Earth very shortly after conditions became favorable for its creation.

Thus the record of life on Earth is at least 3.5 billion years long. It is possible to roughly date the times at which bacteria, and then more complex single-celled organisms made up of cells with nuclei, appeared. Early multicellular life forms are found in the fossil record and so are later more complex ones. The appearance of aquatic life forms and subsequently of the first land animals can be dated. We can trace the evolution of reptiles and then the evolution of some of these reptiles into mammal-like creatures. It is possible to see how many of the plants and animals around us today evolved into their present forms. For example, we know that whales evolved from hoofed mammals, and fossils of intermediate forms—whales with legs—have been found. Even though there are numerous gaps in the fossil record, it is possible to trace the ancestry of many of the organisms that exist on Earth today. Paleontology—the study of fossils—provides an enormous quantity of evidence about the manner in which evolution has proceeded.

Interpretation of the Fossil Record

Fossils have not always been interpreted as providing us with evidence about the past evolution of life. The ancient Greeks, for example, believed that large fossil bones were relics of a race of human giants that had supposedly once existed. When fossils of sea-dwelling organisms were found in rocks far inland or hundreds of feet above sea level, it was often concluded that their formation in the rocks was similar to that of crystals. It was commonly believed that fossils were younger than the rocks in which they were found, and hence there was no reason to think that they furnished a record of the history of life.

Around 1500, Leonardo da Vinci concluded that fossil shells found in Italy were the remnants of ancient oceanic creatures. But Leonardo, as usual, was ahead of his time. A more common belief was that these fossils had been carried inland by the biblical Deluge. In opposition to this view, Leonardo argued that clams could not have been carried 250 miles inland, from the sea to Monferrato in the biblical forty days. Furthermore, he saw that fossils embedded in different layers of sedimentary rock must have been formed at different times. They could not all have been placed there as the result of a single catastrophic event.

But Leonardo's arguments had no effect on current opinion. They couldn't have. He wrote his observations in a notebook, now known as the Leicester Codex, that did not come to light until the 1690s, more than 170 years after Leonardo's death. (This notebook, by the way, is now owned by Bill Gates of Microsoft. To obtain it, Gates outbid several European governments in an auction held at Christie's in 1994.)

By the latter part of the seventeenth century, many geologists had come to believe that fossils were the remains of once-living organisms and that their placement in different layers of sedimentary rock indicated that they were deposited at different times. Sedimentary rock is often formed from tiny pieces or rocks and minerals that have been transported by grav-ity, wind, running water, or glaciers and then deposited over existing rock or sediment. Over time, as more sediment piles up, the deposits are cemented together. When this happens, what was once a loose sediment hardens into rock. Certain other sedimentary rocks are formed when materials that have been dissolved in water precipitate and are deposited at the bottom of a body of water. Gypsum and some kinds of limestone are formed in this manner.

Sedimentary rock is found to have numerous different layers, or strata, with discontinuities between them. The texture and composition of a sin-gle layer has a tendency to be uniform, but different strata will generally not have identical qualities. This is only reasonable because different conditions exist at different times. Since the rock is formed layer by layer, the underlying layers must be older than those above it.

The idea that different layers of sediment have different ages, and that the topmost layers are the youngest was first enunciated by the Danish geologist Nicholas Steno in 1669. If this is true, sedimentary rock must therefore provide us with a chronological history of Earth. Of course Steno had no idea how old the lower strata were. At that time, there was simply no way to estimate their ages. No one knew how long it took for layers of sediments to be deposited, but it was known that the process was slow. And of course it was impossible to tell how much time, if any, might have elapsed between the deposition of different strata.

Steno himself felt constrained by religious doctrine to place all the observed strata within a 6,000-year span. He probably did this willingly, since he was not an irreligious man. In fact, he soon abandoned science for religion, taking holy orders in 1675, and became a bishop in 1677. Since the literal truth of the Old Testament was not widely doubted, this view prevailed until the middle of the following century.

One of the first individuals to question the literal interpretation of the six days of Creation was the French naturalist Georges-Louis Leclerc, comte de Buffon. In 1749, Buffon suggested that the history of Earth could be divided into six different epochs and that Earth had an age of some 75,000 years. This was a huge underestimate. Today we know that Earth is actually 4.6 billion years old. However, at the time, Buffon's ideas were revolutionary. They were very influential as well. Since Buffon's work was widely read throughout Europe, many individuals became convinced that the geological history of Earth could not be confined within a 6,000-year time frame.

Much of what Buffon wrote about geological epochs was speculative. However, even as his ideas were being debated, geologists in England,

Germany, Scandinavia, Italy, and Switzerland were studying sedimentary rock formations. They were able to see an order in the succession of different strata, and they published geological maps and cross sections. By doing so, they prepared the way for scientists who would later study the fossils that were found in strata of different ages.

During the nineteenth century, study of sedimentary rock strata intensified, and it was found that rock strata found in one part of Europe could be matched up with those found in another because they contained the same kinds of fossils. Furthermore, strata bearing certain types of fossils could be matched up with strata a thousand or more miles away. It was even possible to establish chronological relationships. For example, it was found that a group of strata named Juras for the Jura mountains of France and Switzerland had been formed at a later time than a group of strata named Trias in Germany but earlier than a third group in France called Cretaceous. Again, the fossil content of the strata allowed scientists to determine this age relationship. During the 1830s, these findings were formalized and a geologic time scale was established. It was still not known exactly how old any of the rock strata were, but their chronological order was established.

Since sedimentary rocks often contain fossils, a chronological order was also established for the history of life on Earth. Fossils found in older rock strata had preceded those that were embedded in layers of sedimentary rock that had been laid down at later times. Since certain strata characteristically contained certain kinds of fossils, scientists had a record of the life on Earth from the present to times far in the past.

Some attempts to date rock strata were made. Some scientists thought that it might be possible to estimate the rates at which sedimentary rock formed and to guess the age of layers of rock by measuring their thickness and the thickness of the strata that lay above them. However, these attempts were hampered by the fact that there was no way of being sure that sedimentation proceeded at the same rate during past periods that it did during the present. In fact, it seemed more likely that sedimentation would proceed more slowly or more quickly, depending on the conditions that existed at the time. And indeed the estimates that were made turned out to be very inaccurate. For example, near the end of the nineteenth century, one British geologist estimated the age of some rock layers at the beginning of a geological period called the Cambrian to be 75 million years old. Today, we know that they are 600 million years old. It seems that his estimate was too low by a factor of 8.

Today, geologists no longer try to use assumptions about sedimentation rates to estimate the ages of rock formations. They don't have to. Physics provides them with methods that are much more accurate. Radioactivity was discovered in 1896, and physicists have been studying it for more than a century. During this time, they have developed methods of dating almost any object that contains radioactive substances. A small amount of radioactivity is found in numerous natural objects, including sedimentary rock. Thus measurements of the quantities of radioactive substances that rock contains allow scientists to determine its age.

Radioactive Dating

A radioactive element is one that emits any of three different kinds of radiation. They were originally named alpha, beta, and gamma after the first three letters of the Greek alphabet, and they were studied extensively during the early years of the twentieth century. It soon became apparent that the first two types of radiation were actually made up of particles. Alpha particles were composed of two neutrons and two protons. They were identical to the nuclei of helium atoms. Beta particles were nothing other than electrons.

When a nucleus emits either an alpha or a beta particle, it is transformed into a nucleus of another type. For example, the radium-226 nucleus disintegrates into radon-222 when an alpha particle is emitted. This process is called radioactive decay. The numbers 226 and 222 designate the total numbers of protons and neutrons in each nucleus. Radium-226 has 88 protons and 138 neutrons, while radon-222 has 86 protons and 136 neutrons. The two elements, incidentally, are quite different. Radium is a heavy metal, while radon is a gas.

Radium-226 has a half-life of 1620 years. This means that half of it decays into radon in that period of time. But it is not necessary to wait 1620 years to measure a half-life of this magnitude. Radium decays into radon continuously. Some of it decays in the first hour, the first minute, even the first second, while other radium nuclei take much longer than 1620 years to disintegrate into radon.

Now suppose we find a rock that contains both radium and radon. If we want to know the age of the rock, all we need do is measure the relative amounts of the two substances. If we find that there are equal quantities of each, we can conclude that the rock is 1620 years old. It is safe to assume that little or no radon was present originally. It is a rare gas, and only tiny quantities are present in the atmosphere.

As it turns out, measuring radium–radon decay is not a very useful method for dating rocks found in geological formations. For one thing, the half-life is very short. Compared to the many millions or billions of years that have passed since fossils and sedimentary rock formations were created, 1620 years is a very small time. This limits the accuracy of the determination.

If we are to use radioactive decay to measure the ages of rocks, we would like to find some radioactive decays with half lives comparable to the ages of the rocks we want to study. Fortunately, a number of such decays exist. For example, uranium-238 decays into lead-206 in 4,470 million years. This decay is not a simple process. In fact, it consists of some fourteen different steps. The uranium decays into thorium-234, the thorium into protactinium-234, and so on. One of the fourteen steps, incidentally, involves the decay of radium-226 into radon-222.

If a rock contains both uranium-238 and lead-206, then its age can be determined. It is not necessary to measure the quantities of any intermediate elements that may be present. Knowing the half-life of the fourteen-step process and the relative quantities of the parent and daughter elements is all that is needed.

Some types of radioactive decay are more useful for dating purposes than others. For example, as zircon crystals form in cooling volcanic rock, they capture uranium but do not capture lead. Thus when such crystals are found, we know that there were no lead atoms in them originally. Hence any lead that is present must have been created by radioactive decay. The possibility that the results may be skewed by lead that was present in the sample originally can be eliminated.

Parent	Daughter	Half-life
Carbon-14	Nitrogen-14	5,730 years
Uranium-235	Lead-207	704 million years
Potassium-40	Argon-40	1,280 million years
Thorium-232	Lead-208	14,010 million years
Rubidium-87	Strontium-87	48,800 million years

The decay of uranium-238 into lead-206 is only one of a number of radioactive decay processes that can be used to date rocks and other materials scientists study. The radioactive decays used for the dating of fossils are shown in the table.

Carbon–nitrogen dating, often called simply "carbon dating," is probably the one that is most familiar to lay people. However, it is not of much use in paleontology, since it can rarely be used to determine the ages of rocks in geological formations. The half-life of carbon-14 is so short that the method becomes inaccurate when attempts are made to determine dates much greater than 50,000 or 60,000 years. But fortunately, there are a number of decay processes that can be used to date older geological formations. Note that the half-lives of the thorium–lead and rubidium–strontium decays are greater than the age of Earth itself, which is only about 4,600 million (4.6 billion) years.

Dating Fossils

Fossils are typically found in sedimentary rock. But sedimentary rock cannot be dated directly. The particles of sediment from which the rock is formed are the products of erosion. Wind, rain, and the action of glaciers cause small particles to be eroded from mountains and other rock formations. Since sediments are formed from preexisting particles, any attempt to date these particles would produce ages greater than the amount of time that has passed since the sediment was laid down.

However, this is not a serious difficulty. Newly formed volcanic rocks are often embedded in sediment, and the dating of these rocks provides us with an age for the layer of sedimentary rock with which we are concerned. Since, as I have noted previously, fossils must be of the same age as the sediment, finding the ages of the fossils becomes a relatively straightforward matter.

Furthermore, there are checks that prevent an erroneous dating of a volcanic rock from skewing results. The same sedimentary layers and the same fossils are found in many different geographical regions. If an age were obtained for a sedimentary layer that was different from the ages obtained for the same layer in different regions, the discrepancy would immediately become apparent.

Consequently, it is possible to establish with a fair degree of certainty that a given dinosaur bone is, say, 85 million years old or that a skull of our predecessor *Homo erectus* does indeed have an age of 1.5 million years. We can say with confidence that the traces of bacteria found in very old rocks are indeed the remnants of organisms that lived 3.5 billion years ago. Thus it is possible to trace the history of life on Earth with a fair degree of accuracy.

So far, I have said nothing about the creation of fossils and have noted only that they are generally found in sedimentary rock. Thus, before I go on, it might be a good idea to discuss fossilization and the types of organisms that are likely to become fossils in some detail.

The Creation of Fossils

When an organism dies, its body normally either is eaten by scavengers or undergoes bacterial decay. It is generally only the hard parts—bones, shells, and teeth—that survive for any length of time. But these hard parts do not endure, either, unless something happens to preserve them. This "something" is normally burial in sediment.

The more quickly the hard parts become buried in sediment, the more likely they are to survive. For example, the shell of a dead aquatic organism normally does not fossilize at all. It is more likely to be broken up by the action of waves and tides. But if it does become buried in sediment, it may remain there for a long period of time. If new layers of sediment begin to form on top of the old ones, then the lower layers will be compressed. Water will be forced out of them, and the individual particles that make up the sediment will be cemented together. When this happens, the hard parts that the lower sediment contains may be destroyed or deformed or they may be preserved as fossils embedded in rock.

Not all fossils come from dead organisms' hard parts. Sometimes the soft parts leave impressions in materials that contain no oxygen, such as volcanic ashes and certain kinds of mud. This process is relatively uncommon. However some fossils of soft-bodied animals have been found. There also exist trace fossils, such as fossilized footprints or burrows. And of course insects are sometimes fossilized when they are preserved in amber. Nevertheless, the vast majority of fossils come from organisms' hard parts.

Sediments are found at the bottom of columns of water—in ocean beds, for example. Consequently, the organisms most likely to leave fossils are those that live at the bottom of an ocean or lake. Swimming organisms are less likely to have their hard parts preserved as fossils, and terrestrial organisms are the least likely of all to be fossilized. It is estimated that only about 20 percent of the organisms living in the oceans today are likely to fossilize easily. The percentage of land organisms likely to become fossils is thus considerably lower.

Only a small number of the species that live in oceans and lakes today are likely to be fossilized. Many of them have no hard parts, and others have life-styles that make it unlikely that they will be preserved in sediment. As we have already seen, terrestrial organisms are less likely to be fossilized than aquatic animals. And of course plants, which are made up of nothing but soft parts, rarely fossilize at all.

Most fossils remain buried deep in sedimentary rock, and remain inaccessible. If a layer of sediment is exposed because of some geological process such as erosion, the fossil soon disintegrates. That is why it is necessary to dig for fossils. They are rarely found lying around on the surface of Earth.

Organisms leave fossils in only a few cases. Many fossils are destroyed by various kinds of geological action. The vast majority of those that are preserved are inaccessible (many remain miles below the surface of Earth) or are never found. Thus we should expect the fossil record of life on Earth to be fragmentary. And indeed it is so fragmentary that the majority of the species living today have no fossil history. Nevertheless, a fossil has the potential to provide scientists with an enormous quantity of information. New discoveries add to our knowledge about the history of life every day. For example, scientists are still learning a great deal more about the dinosaurs, which died out some 65 million years ago. And dinosaur fossils constitute only one of thousands of different types of fossil that have been studied.

A Short History of Life

Earth was formed some 4.6 billion years ago from a disk of dust and gas that surrounded the sun. Parts of the disk that were denser than average exerted gravitational force on their surroundings, and clumps of compacted dust were formed. As these clumps grew larger, the gravitational attraction they exerted on other material became larger yet. Soon bodies the size of asteroids were orbiting the sun. Collisions between these asteroids caused them to clump together, creating bodies that were larger yet. Over a period of about 100 million years, one of these bodies grew to the size of a small planet. The nascent Earth continued to experience collisions. In fact, there is reason to think that it may have once collided with a body the size of Mars, and that this collision separated what is now the moon from Earth. The collisions that Earth experienced during its early history were continuous and intense. And they generated so much heat that the surface of Earth remained molten. Earth was not yet a place where life could evolve.

However, by approximately 3.8 billion years ago, Earth had cooled sufficiently that a solid crust could form. Originally, there was no water on Earth's surface. But then volcanic eruptions released water from Earth's mantle, and oceans formed. Almost as soon as they came into existence, the oceans were full of organic chemicals that were to be the building blocks of life. The solar system is full of such chemicals, including amino acids—the components of proteins—and they would have been carried to Earth's surface in falling interplanetary dust particles. Passing comets would also have deposited organic chemicals on Earth as well.

It is not known exactly how life began. Several plausible schemes have been suggested. However, we can't travel back 3.5 billion years in time to see which idea is correct. But it is probably safe to assume that some kind of chemical evolution preceded the creation of life. Under the proper conditions, systems of very complex organic chemicals will form. At some point, one such system of chemicals must have evolved into the first living cell.

Though scientists can only make guesses about the origin of life, they are on much firmer ground when they speak of what primordial life is like. Life leaves traces in the geological record. It was already abundant enough 3.5 billion years ago that fossil bacteria can be found in rocks of that age. Since Earth's surface was too hot to support life until around 3.8 billion years ago, the first living organisms must have come into existence fairly quickly. Those first organisms may not have any descendants alive today. It is conceivable that life was wiped out several times, only to evolve anew. Nowadays, the extinction of the dinosaurs is attributed to the collision of Earth with an asteroid some 65 million years ago. It is certainly conceivable that the frequent collisions that primordial Earth experienced may have caused total extinction one or more times.

But by about 3.5 billion years ago, life had gained enough of a foothold that it was able to endure. There is every reason to think that the bacteria that are found in ancient rocks in Australia and Africa are indeed our ancestors and the ancestors of every living organism that exists today. We are all descended from bacteria.

Primordial Earth was swarming with bacteria. It still is today. Bacteria are by far the most common kind of organism. Their numbers far exceed those of other kinds of organisms, and they live in every conceivable habitat. There are about 100,000 bacteria on every square centimeter of human skin, and one small spoonful of soil may contain 10 trillion bacteria. They are found in ocean vents where water flows upward from Earth's interior at temperatures as high as 480°F. They are found in pools on glaciers and on the surfaces of burning coals. They live in rocks thousands of feet below Earth's surface and in sediments at the bottom of the oceans. It may even be that the biomass of bacteria exceeds that of all other living organisms combined. One estimate puts the total mass of underground bacteria (the most numerous) at 200 trillion tons.

During the first 1.5 billion years of life, only bacteria populated Earth. Then, a major step in evolution took place. More complex organisms, called eukaryotes (pronounced "you-carry-oats"), began to be fossilized about 2 billion years ago. All the plants and animals that exist today are composed of eukaryotic cells. Fungi, modern single-celled animals (such as amoebas), and algae are also eukaryotic.

Bacteria are relatively simple in structure. A bacterial cell contains no nucleus or other complex structures. A eukaryotic cell, on the other hand, is more complex. The cell's DNA is confined within a nucleus, which is enclosed by its own membrane, and there are numerous other structures, called organelles, within the cell. Some eleven different organelles are found in eukaryotic cells but not in bacteria. I won't attempt to name them all, but two kinds are worthy of special mention. Both animal and plant cells contain mitochondria, which convert energy-rich food particles into energy that can be used by the cell. Plants also contain chloroplasts, which carry out photosynthesis.

Here it will be necessary to say something about the function of DNA. To interrupt my discussion of mitochondria and chloroplasts as little as possible, I will keep it short and simple. The DNA double helix is really nothing more than a complex code that tells a cell what kinds of proteins to make. Cells are made of proteins, and enzymes—which act on proteins—are themselves made of proteins. Small chunks of DNA code for the production of the various amino acids that are the components of proteins. In other words, DNA is a kind of blueprint for making cells and for making organisms, which are large collections of cells.

Both mitochondria and chloroplasts contain their own DNA, and they reproduce independently of the cells that enclose them. Mitochondria—but not chloroplasts—have a genetic code that is slightly different from the "normal" DNA in the nucleus. Chunks of mitochondrial DNA sometimes code for the production of different amino acids (remember that amino acids are the components of proteins) from identical pieces of nuclear DNA. This is somewhat surprising, since the genetic code that is found in nuclear DNA is shared by all living organisms.

It is not known exactly when eukaryotes began to evolve. In 1999, the Australian scientists Jochen J. Brocks, Graham A. Logan, Roger Buick, and Roger E. Summons reported that they had found fossils of biological lipids in 2.7-billion-year-old shale in Australia's Pilbara Craton. Lipids are chemicals that are components of eukaryotic cell membranes and the membranes of such organelles as mitochondria. The first unmistakably eukaryotic fossils appear only 2 billion years ago, evidence that eukaryotes may have begun to evolve sooner than was previously thought.

When bacterial fossils are examined, scientists can determine what they looked like only on the outside. It is often impossible to infer anything about the bacteria's internal structure. However, it is thought that the predecessors of modern eukaryotic cells began to evolve at least 3 billion years ago from primitive nucleated cells that had ingested smaller bacteria. Over a period of a billion more years, stable relationships developed between the host cells and their symbionts, and the ingested cells began to lose functions that were no longer needed in their new environment. Over time the latter evolved into the chloroplasts and mitochondria observed in cells today.

Different kinds of evidence support the idea that evolution proceeded in this manner. First, chloroplasts and mitochondria have their own DNA. Furthermore, protein synthesis in these organelles exhibits numerous similarities to protein synthesis in bacteria. Finally, it is possible to observe similar symbiotic relationships today. For example, certain marine slugs and related animals feed on green algae. The chloroplasts in the cells of the algae continue to carry out photosynthesis for some time after they are incorporated into the slugs' cells, and the carbohydrates produced by photosynthesis are an important source of nutrients for the slugs. The fact that chloroplasts can be incorporated into animal cells and continue to function gives support to the idea that they formed symbiotic relationships in the past. If they can become symbionts of animals today, this is evidence that they are adapted to this role.

There are many other types of organelles besides chloroplasts and mitochondria. However, they do not contain their own DNA, and little is known about their evolution. It is conceivable that they too evolved from ingested symbionts. Alternatively, they may be structures that evolved inside primitive eukaryotic cells. Though scientists understand the broad outlines of the evolution of eukaryotic cells, they are not able to fill in all the details.

An important event associated with the evolution of the eukaryotes was the presence of significant quantities of oxygen in Earth's atmosphere. The atmosphere contained no oxygen originally. It was probably composed primarily of carbon dioxide and nitrogen. But as photosynthesizing single-celled organisms evolved, they began to release oxygen. Significant quantities of this gas were present around 2 billion years ago, and by 1.5 billion years ago oxygen levels were close to those that exist today. The most likely reason for this abundance was that microorganisms were producing oxygen at a greater rate than they had previously. There may have been photosynthesizing bacteria as long as 3.5 billion years ago, but photosynthesis began to proceed at a rapid rate only around the time that the eukaryotes evolved. Possibly the chloroplasts in eukaryotic cells produced oxygen more efficiently than did their bacterial predecessors.

Oxygen was a poison to most of the organisms that existed 1.5 billion years ago. Hence the increase in oxygen levels was an ecological catastrophe. However, organisms evolved that were more oxygen-tolerant, and it is from them that most living organisms are descended.

Multicellular Life

Complex multicellular animals might never have evolved if oxygen levels had not become high. A substance known as collagen is an important connective material in such animals. But collagen cannot be manufactured without oxygen. Oxygen is also required if structures such as gills and circulatory systems are to develop. Thus the early eukaryotes, if they were indeed responsible for the increased oxygen levels, prepared the way for life that was even more complex.

The earliest fossils of multicellular animals come from the Ediacarian deposits in Australia and from similar deposits elsewhere in the world, which date from a period 670 to 550 million years ago. The Ediacarian fossils are the remains of soft-bodied marine animals such as worms and jellyfish. In many cases, the fossils consist of little more than markings left by burrowing worms. As we saw previously, soft-bodied animals rarely fossilize. Thus the record of early multicellular life is somewhat fragmentary. Most, and possibly all, of these animals became extinct around 550 million years ago.

And then, suddenly, around 535 or 530 million years ago, a very large number of animals appear in the fossil record. By this time, marine organisms (there was not yet any life on land) had evolved hard parts, which were more likely to be preserved in the fossil record. So many different kinds of life came into existence at this time that scientists sometimes speak of the Cambrian Explosion.

This event is called "Cambrian" because it occurred during the geological period of that name. The Cambrian period lasted from around 530 million years ago to about 505 million years ago. During this time, an astonishing number of different kinds of marine animals evolved. Some of these are so different from anything alive now that they look very bizarre to the human eye. One animal, for example, had five eyes and a frontal nozzle that somewhat resembled a vacuum cleaner. Another was a stalked animal that looked something like a flower and had a mouth adjacent to its anus. Other animals are more familiar looking and are probably the ancestors of organisms living today.

It is not very difficult to image how multicellular life evolved. Single-celled organisms often form colonies, and some of these colonies must have gradually evolved into true multicellular life. The plausibility of this idea is enhanced by the fact that there exist organisms today that are colonies of single-celled animals but that nevertheless exhibit characteristics of multicellular organisms. For example, the Portuguese man-of-war is a jellyfishlike animal (it isn't a true jellyfish) that consists of hundreds of different single-celled organisms that are genetically identical. These individuals assume many different shapes and functions. For example, one of them becomes the float. Some become specialized for capturing prey, and others become male and female sexual organs. However, there is no brain and nervous system that sends messages to the individual organisms that make up the jellyfish. Each of them retains its own identity.

Evolution Since the Cambrian Era

Beginning with the Cambrian period, the fossil record becomes richer. There are more animals with hard parts. At first these hard parts were not bones. Bones had not yet evolved. However, numerous animals had shells and skeletons. Here I am referring to exoskeletons, such as those possessed by the modern crabs and lobsters. Vertebrates—animals with spinal columns—had not yet appeared.

Land was first colonized by microbes, colonies of bacteria that formed mats near the water's edge around 1.4 billion years ago. Scientists are not entirely certain what kinds of bacteria they were. The fossil evidence for their existence is chemical. Localized areas rich in organic chemicals are found in the fossil record. These chemicals must have been produced by microbes of some kind. The best guess seems to be that they were cyanobacteria, which are sometimes called blue-green algae.

The first terrestrial plants appear during the geological period called the Ordovician, which lasted from 505 million years to 438 years ago.

Animals followed; they began to colonize the land during the Devonian period, which lasted from 408 million years to 360 million years before the present. One should not draw the conclusion, by the way, that these animals were somehow more "evolved" than those that remained in the oceans. They were simply a separate evolutionary line, and life that remained in the sea continued to evolve just as terrestrial life did. Mammals are not "higher" on the evolutionary scale than reptiles, and reptiles are not more "evolved" than fish. Since the idea that evolution is some sort of ascending ladder is one of the most common misconceptions about the subject, this point cannot be emphasized too much. A frog is just as "evolved" as a human being. Both are the result of 3.5 billion years of evolution.

Reptiles and Mammals

Human beings are mammals, so perhaps it is not surprising that they are especially interested in mammal evolution. To be sure, paleontologists can become quite excited by interesting new discoveries about snails or about such long-extinct animals as the trilobite. But most nonscientists do not. So perhaps I will be somewhat justified in paying special attention to the evolution of mammals from mammal-like reptiles.

The first mammalian fossils were formed a little less than 200 million years ago. These early mammal fossils constitute a group with the mouth-filling name morganucodontids. Members of this group had mammalian jaws and teeth and a mammalian gait and were probably also warm-blooded. The ancestry of these mammals can be traced back through a group called the mammal-like reptiles, which gradually evolved over a period of some 100 million years. When the fossils of these animals are examined, a clear progression can be seen, especially in the gait and in the jaw and tooth structure.

Around 300 million years ago, lizardlike animals called pelycosaurs appeared. This animal had a feature that set it apart from other reptiles: an opening in the bones behind the eye. A muscle that passed through this opening closed the jaw. This is the first sign of a powerful mammal-like bite that is uncharacteristic of reptiles.

Pelycosaurs evolved into three main groups over a time span of about 50 million years. Then, suddenly, most of them went extinct about 260 million years ago. But a few survived and gave rise to a group of animals known as the theraspids, which were even more mammal-like. One subgroup of the theraspids, called the cynodonts, developed structures that were even more mammal-like. Finally, true mammals appeared.

It is not necessary to remember any of these Latin-derived names. I am using them only to distinguish one kind of animal from another. The important point is that we can see a clear progression over a period of 100 million years. This progression does not imply that there was a single line of evolution. On the contrary, the mammal-like reptiles split into a number of different evolutionary lines on a number of different occasions. Most of the species in these different lines eventually went extinct. Those that survived evolved into mammals.

The early mammals were not the dominant form of terrestrial life that they are today. In fact, they have been called "the rats of the age of dinosaurs." None were much larger than a ferret, and they occupied the ecological nooks and crannies of the reptilian world. Only after the dinosaurs became extinct 65 million years ago did they have the opportunity to expand into the ecological niches occupied by their predecessors. It should not be imagined that they survived because they were somehow better adapted than the dinosaurs. The dinosaurs were superbly adapted to their environment and ruled Earth for nearly 200 million years. The mammals have been dominant for only 65 million; they seem to have assumed this role by default.

In other words, the mammals are not more "evolved" than their dinosaur predecessors; they owe their success to a chance encounter of Earth with an asteroid 65 million years ago.

Human Evolution

Self-centered creatures that we are, we pay the greatest amount of attention to our own evolution. Like monkeys, apes, lemurs, and tarsiers, we are primates. Our closest living relative is the chimpanzee. Humans and chimpanzees are genetically very close. They share about 98.5 percent of their DNA. But we are not, of course, descended from chimpanzees or from any other living ape. The human and ape lines diverged about 5 million years ago. In other words, humans and apes have a common ancestor, and both have been evolving for 5 million years since the line split.

Modern human beings are the only living member of the family Hominidae, commonly referred to as the "hominids." Our predecessors *Homo habilis* and *Homo erectus* were hominids, as were the australopithecines, animals that lived on the African savanna from about 5 million to 2 million years ago. The australopithecines walked upright as we do (by comparison, the gorilla and the chimpanzee are knuckle walkers, not fully upright), and they had brains about the size of those of chimpanzees.

Before I go any further, it might be well to mention that evolutionary "trees" typically bear a greater resemblance to great bushes than they do to long, tall saplings. We normally do not find that evolution proceeds in a single line. One species does not evolve into another without producing offshoots. Most often, the evolutionary tree contains numerous different branches that sprout in different directions. Typically, a variety of different species evolves, and many of them become extinct.

Thus, if a fossil hominid that preceded us is discovered, it does not necessarily follow that this hominid was one of our ancestors. During various times in our evolutionary past a number of different hominid species coexisted. In each case, only one of them was part of the evolutionary line that led to *Homo sapiens*.

There were at least four species of australopithecine. They are *Australopithecus afarensis, Australopithecus africanus, Australopithecus robustus,* and *Australopithecus bosei.* It has been suggested that two other distinct species, *Australopithecus aethiopicus* and *Australopithecus crassidens,* also existed. *Australopithecus afarensis* is believed to be ancestral to the others and to human beings. But *Australopithecus robustus* and *Australopithecus bosei* are not our ancestors. On the contrary, they are species that coexisted with the first member of the genus *Homo, Homo habilis,* which evolved about 2 million years ago. *Homo habilis* used tools, such as flaked stone artifacts.

Homo habilis had a cranial capacity of some 500 to 800 cubic centimeters, considerably smaller than that of modern humans who have brains that measure about 1,450 cubic centimeters. In fact, the average is only slightly greater than that of a chimpanzee, which has a brain that measures about 600 cubic centimeters. Although *habilis* used tools, there is some disagreement as to whether the members of this species communicated verbally. Perhaps this is something we will never know. Speech, after all, does not fossilize. Since we have only the bones to work with, it will always be necessary to make inferences about such matters.

There is some disagreement among paleontologists as to whether the fossil remains of *Homo habilis* represent one species or two. However, it is agreed that at least one species (if there were indeed two) was ancestral to *Homo erectus*. *Homo erectus* appeared about 1.6 million years ago and existed for a much longer time. *Homo habilis* seems to have been a transitional stage between *Homo erectus* and the australopithecines.

Homo erectus fossils are found in many parts of the world. Indeed, the first *Homo erectus* fossil found was discovered in Asia. Thus this hominid ranged much more widely than its predecessors, which appear to have lived only in Africa. *Homo erectus* had an average cranial capacity of about 1,000 cubic centimeters. It was the first hominid to use fire and to live in caves.

Homo erectus lived from about 1.6 million years ago until around 250,000 years ago. This hominid is followed in the fossil record by *Homo sapiens,* which evolved about 250,000 years ago. It is difficult to pin down the exact time that *Homo sapiens* first appeared. *Homo erectus* had a wide geographical range, and different populations evolved at different rates. This created a pattern called mosaic evolution. Fossils that are intermediate between *Homo erectus* and *Homo sapiens* have been found in a number of different locations, but the pattern of evolution is not always clear. It is thought, however, that the first members of the species *Homo sapiens* lived in Africa.

For some time, the status of Neanderthal man, which lived during the period from 100,000 years ago to about 35,000 years ago, was a matter of controversy. Some thought that Neanderthal was a form of *Homo sapiens* that was adapted to cold climates and named it *Homo sapiens neanderthalis* (in this classification scheme, we would be *Homo sapiens sapiens*). However it is now agreed that Neanderthal was a distinct species, *Homo neanderthalis,* which was contemporary with *Homo sapiens.*

Neanderthal had a brain that was slightly larger than that of *Homo sapiens*. However, Neanderthal brains were organized somewhat differently, and there is no reason to think that Neanderthal necessarily equaled *Homo sapiens* in intelligence. Neanderthal may have interbred with *Homo sapiens* to some extent, so it is not certain whether this species died out or was simply absorbed into the human population.

There is some uncertainly about the origins of Neanderthal. The fossil record of the period lasting from 400,000 to 100,000 years ago is rather sparse. Some skull fragments of Neanderthal's ancestors have been discovered in Europe, but they are more like *Homo sapiens* than they are like *Homo erectus,* from which Neanderthal presumably evolved. There are real gaps in the fossil record, periods during which few fossils were preserved.

Microevolution

The fossil record tells us how evolution has taken place over the last 3.5 billion years. As we have seen, some of the changes that have taken place are dramatic indeed. If these changes, which happened over long periods of time, were indeed the result of evolution, then it ought to be possible to see small changes in the present-day world.

As a matter of fact, numerous small evolutionary changes have been observed. For example, the European house sparrow, which was introduced into North America in 1852 and has since spread across the entire North American continent, has exhibited observable evolutionary changes. As the species spread, it underwent noticeable changes. Furthermore, different degrees of change are seen in birds that live in different habitats. For example, sparrows that live in or near Edmonton in the Canadian province of Alberta have a longer wing span than those that live in Death Valley, and those that live in Death Valley are lighter in color than those that inhabit Mexico City.

The difference in color between house sparrows in different geographic locations is immediately obvious to those who are familiar with the birds. In fact, in Hawaii, the sparrows are so differently colored that bird watchers from other parts of the United States do not immediately recognize them as sparrows.

And yet all these different varieties are descended from a single breeding population that was established in Brooklyn in 1852. In a little over a century, birds that originally looked very much alike have evolved into a large number of different varieties. Of course, no new species of house sparrow has arisen. One would not expect that to happen in so short a time. However, there has been enough change to show that evolution is very real, and that it is something that is happening constantly.

The best-known story of evolutionary change is that of the peppered moth. Specimens that were collected in Great Britain before the industrial revolution always had a gray "peppered" appearance. A dark form of the moth was first seen near Manchester in 1848. The dark form then became more common as soot killed lichen on the bark of trees, causing the trees to become darker in color. Eventually, more than 90 percent of the moths in polluted areas were seen to be dark in color, while the light form remained common in areas where there was little or no pollution.

Clearly, evolution had taken place. But why? It seemed natural to assume that the dark-colored moths would be better camouflaged on darkened trees and would therefore be less likely to be eaten by birds. Indeed, early studies seemed to confirm this. When the moths were placed on tree trunks, the more conspicuous light-colored moths were more likely to be eaten. However, these studies were somewhat flawed. The moths ordinarily do not alight on tree trunks; they prefer the branches of trees.

However, later studies remedied this defect. In 1973, the British biologist H. B. D. Kettelwell performed an experiment in which he released dark- and light-colored moths in unpolluted and polluted countrysides and later recaptured some of them, using mercury vapor lamps to attract them. He found that the light variety survived considerably better in unpolluted woodland, while the dark variety had a survival rate that was almost twice as high when released in areas that were polluted. Since the survival rate was the quantity that was measured, no uncertainty was introduced by placing moths in trees in a manner in which they would not ordinarily alight.

But predation by birds is probably not the whole story. A study carried out in 1980 seemed to show that the light-colored form of the moth had a survival rate about 30 percent below that of the dark-colored form. And dark forms of other species, including beetles, pigeons, and cats, have become more common in polluted areas even though they are not preyed upon by birds. So bird predation may not be the only thing that causes one form of the moth to be more likely to survive than the other. The genetic mutation that caused the dark forms to appear may be advantageous even when there are no predators around. It seems that one must conclude that, although evolution has occurred, more than one factor may be at work in causing one form to be more likely to survive and reproduce than the other.

Human activities are responsible for much of the microevolution that takes place in the world today. One obvious example is bacterial resistance to antibiotics. The more these drugs have been used, the more resistant certain kinds of bacteria have become. Pesticide resistance in insects is another example. For example, DDT was first used widely in the 1940s to combat malaria, which is carried by mosquitoes. In India, DDT remained effective for about ten years before DDT-resistant mosquitoes began to appear. Malaria-causing mosquitoes now become resistant to DDT within months, rather than years, when they are first sprayed with it. Since the gene for resistance was now present in the population, it could spread rapidly once this pesticide was used. The mosquitoes that have the gene for resistance tend to survive and produce offspring, while the nonresistant mosquitoes perish. This has had unfortunate effects. Although DDT use caused the number of worldwide cases of malaria to fall to around 75 million per year in the early 1960s, the disease now claims about 300 million victims annually.

When an insect becomes resistant to one insecticide, it is often sprayed with another. In many cases, the same evolutionary pattern repeats itself, this time on a shorter time scale. For example, the Colorado potato beetle, which evolved DDT resistance in seven years, became resistant to three other pesticides with which it was sprayed in five years, two years, and two years, respectively. This is a fairly common pattern. It appears that the genes that cause resistance to one pesticide also help insects to become resistant to others.

New Species

One sometimes hears it said that, although small evolutionary changes are seen in nature, the creation of a new species has never been observed. This is not true. The evolution of a new plant species has been seen in nature, and numerous other new species have been created by scientists.

The most common definition of the term "species" is this: a population that is reproductively isolated from other, related species. If two populations do not interbreed or do not produce fertile offspring when they do, they are said to be distinct species.

The species may be reproductively isolated for any of several different reasons. Some are so different from one another that they could not produce offspring if they did mate. For example, no one has never seen a cross between a hippopotamus and a giraffe, and no one ever will. The two are genetically too unalike to produce offspring. In other cases, breeding does not take place because two species with a common evolutionary ancestor have evolved different courtship rituals; members of the two species simply do not recognize one another as potential mates. Finally, some pairs of species produce offspring that are sterile. The most familiar example of this is the mule, which is the offspring of a jack donkey and a mare. And of course the less common hinny, a cross between a stallion and a female donkey, is sterile too.

Scientists have gained some insight into the manner in which reproductive isolation might come about by studying the apple maggot fly, *Rhagolettis pomenella*. Agricultural records show that a strain of this fly began infesting apples in the 1860s. Formerly it infested only hawthorn fruit. The two strains are now behaviorally isolated. Flies that infest one kind of plant do not breed with the flies that live on the other. Furthermore, it has been shown that the two have developed genetic differences from one another.

Are there now two species of fly where there was only one before, or are they just two different strains of a single species? This is not an easy question to answer. The divergence of one species into two is a gradual process, and defining the point at which they become two separate species is a somewhat arbitrary procedure. It is something like trying to decide at what exact point a single-celled organisms that is dividing in two becomes two individuals rather than one.

However, there are other cases in which we can definitely say that new species have evolved. For example, two new species have evolved in the plant genus *Tragopogan* within the last fifty or sixty years. They are called *Tragopogan mirus* and *Tragopogan miscellus*. They were created when one species fertilized another, producing hybrids. In each case, the hybrid could not fertilize or be fertilized by either of its two parent species. The new plants were reproductively isolated from other similar plants. Recall that reproductive isolation *defines* the term "species."

The species I have described arose spontaneously, without any intervention by scientists. It stands to reason that if we can observe natural speciation, we should be able to create new species in the laboratory. This is indeed the case. In fact, the creation of new plant species by hybridization is commonly carried out for agricultural purposes. For example, many of the grains we eat did not evolve naturally. They represent species that did not exist in nature until humans intervened.

Plant breeders commonly create new species. Most of the irises, dahlias, and tulips that we grow in are gardens are members of artificially created species. So are orchids. It is estimated that about 300 new hybrid species of orchid are created every month.

Plants are genetically more pliable than animals, and it is consequently much easier to create new species. For example, there are numerous different breeds of dog, but no dog breeder has ever created a new species. A dachshund and a great Dane could easily produce fertile offspring if they were so inclined. In fact, the breeding of domestic animals must continue generation after generation if they are not to revert to a wild form. For example, after a few generations, feral pigs will take on a somewhat different appearance and will even grow tusks.

Since it is much more difficult to create new animal species than new species of plants, it may be centuries before scientists see any new animal species appear naturally. However, the fact that a few new plant species have arisen naturally, while thousands of others have been created by scientists and by breeders, is yet another piece of evidence that confirms that evolution is an ongoing process in the natural world.

Frogs' Legs and Bats' Wings

The creation of new species by hybridization—either natural or artificial—of plants is quite common. It has been estimated that 50 to 80 percent of all flowering plant species are hybrids. However new animal species are more likely to arise when one species splits into two different strains and these strains then evolve into separate species. If this is indeed the case, then we should expect to see similarities between species that have a common ancestor. One example of this has already been pointed out. Humans and chimpanzees have genomes that are 98.5 percent identical.

It is also possible to see similarities between species whose ancestors diverged into separate species many, many millions of years ago. In fact, all mammals share a common body plan. For example, there is a one-to-one correspondence between the bones of a human hand, the bones in the front flipper of a whale, and the bones of the wing of a bat. The same correspondence is seen in all tetrapods, including birds, amphibians, and reptiles (a tetrapod is an animal with four limbs; a horse is a tetrapod, and so is a bird, which has two legs and two wings). The same bones are seen in the leg of a frog, the wing of a bat and the arm and hand of a human being.

All tetrapods have limbs with five digits. In some species, such as the horse, some of the digits disappear in the adult form; however they can still be seen in an embryonic horse. There is no reason why the number would always be five if these animals did not have a common ancestor. In many of them a three- or four- or seven-digit limb would be just as functional.

Such correspondences are called homologies. The existence of a five-digit limb in a wide variety of different animals is just one of many different examples. There are numerous others. For example, the wings of birds are all constructed according to the same basic plan. Thus we can conclude that birds have a common ancestor. On the other hand, the wings of bats, though they are also used for flying, are constructed differently—the bones are arranged in different patterns. Thus we can conclude that birds and bats do *not* have a recent common ancestor.

There is another kind of homology that provides evidence that all organisms have a common ancestor. This is the fact that they all use the same genetic code. Although many different kinds of genetic codes are theoretically possible, all organisms use the same one. The DNA of a bacterium, for example, codes for the creation of proteins in the same way mammal DNA does. This fact has been confirmed by experiment. Genetic material from a rabbit can be injected into a bacterium, where it is observed to function in the same way. Rabbits and bacteria may not look very much alike, but they are homologous on the molecular level.

The existence of homologies does not provide direct evidence for evolution in the way that observations of microevolution and species creation do. It is strong evidence nevertheless, because we see exactly the kind of homologies that would be expected. The bones of a human hand or in the limb of some other mammal are like those in the wing of a bird but not like any bones found in a fish. But the fossil record indicates that both birds and mammals evolved from reptiles. There should be more similarity between them than between a mammal and a fish. The evidence indicates that some fish evolved into amphibians, and that some amphibians evolved into reptiles. Ancient fish are thus more distant ancestors than reptiles are and should therefore be more unlike us.

Evolution Is a Fact

The idea that life on Earth has been evolving for some 3.5 billion years is not a theory. It is a fact. It is the only possible interpretation of the numerous different kinds of evidence that scientists have been uncovering for well over a century. In this chapter, I have cited only some of the different kinds of evidence for evolution. A book that discussed them all in detail could probably not be written. At the very least it would be made up of hundreds of volumes. Paleontologists, zoologists, botanists, geneticists, ecologists, embryologists, and other scientists unearth new evidence for evolution every day.

One might ask why we still speak of "Darwin's theory of evolution" if this is indeed true. I think that this question is easily answered. In the first place, the word "theory" is not used the same way in science as it is in everyday life. A scientific theory is an organized body of knowledge, usually of knowledge that is well confirmed. It does not refer to something hypothetical, as the colloquial term "theory" might. For example, Newton's laws of motion constitute a theory, one that has been confirmed over and over again during a period of centuries. There is nothing hypothetical about Newton's system of mechanics. It has, to some extent, been superseded by Einstein's relativity, but it is still valid within its own domain. On the other hand, a lawyer's "theory" about the manner in which a crime was committed that is given to a jury is something more tentative, and it is up to the jury to decide whether the idea seems plausible or not.

Furthermore, as I will explain in the next chapter, Darwin's "theory of evolution" is not a single theory but rather a combination of four or five. In his book *On the Origin of Species,* which was published in 1859, not only did Darwin attempt to show that evolution had taken place, he also theorized about the mechanisms that caused species to evolve. Showing that something has happened and attempting to understand precisely *how* it happened are two different things.

Today, the evidence for evolution is considered to be overwhelming, and Darwin's theories about how it happened have won near-universal acceptance. Nevertheless, it is still possible to argue about some of the details. A description of some of the arguments that are currently taking place will have to be put off until later. It will be necessary to look first at what Darwin really said.

Darwin's Five Theories of Evolution

S ome tales are too good to be entirely true.

For example, there is a famous story about an exchange that took place in June 1860 between the English biologist T. H. Huxley and the Anglican bishop Samuel Wilberforce at the annual meeting of the British Association for the Advancement of Science. Darwin's book *On the Origin of Species* had been published the previous year, and evolution was one of the topics that was to be debated. Darwin, who avoided controversy whenever he could, stayed away from this meeting. However, it was attended by a number of Darwin's supporters.

Darwin's theory came under attack out the outset. On Thursday, June 28, it was attacked by the British anatomist and paleontologist Richard Owen. One of Owen's main arguments against evolution was the supposed difference between the human brain and the brain of the gorilla. The gorilla's brain, Owen said, "presented more differences, as compared with the brain of man, than it did with the brains of the very lowest and most problematical of the Quadrumana." If this claim had been true, it would have set human beings apart from the other mammals and would have implied that humans were not the product of evolution. So Huxley rose to Darwin's defense. He flatly contradicted Owen on this point, promising to reply in greater detail later in print.

The topic of evolution did not come up on the following day, Friday. But it wasn't long before the argument was resumed with a vengeance. On Saturday, June 30, Wilberforce, who was then Bishop of Oxford, was scheduled to speak. At first, Huxley was inclined to stay away from this session and return home instead. He apparently had no desire to debate with Wilberforce, telling the Scottish publisher and author Robert Chambers that he did not wish to speak and find himself "episcopally pounded." Wilberforce was well known as an orator and had acquired the nickname "Soapy Sam" for his eloquence. Huxley must have felt that his debating skills would prove to be inferior to those of Wilberforce.

But in the end, Huxley changed his mind and attended after all. When he arrived, he found that over 700 people had assembled for the session, quite a large number for a meeting that was closed to the public. A number of speakers, both scientists and clergymen, spoke, and then Wilberforce took the floor. Wilberforce spoke in the persuasive manner for which he was known and marshaled a number of arguments against Darwin's theory of evolution. For the most part, these arguments were not his own; he had been coached beforehand by Owen.

At the end of his oration, Wilberforce turned to the audience, asking whether they supposed they were really derived from beasts. Then he turned to Huxley, gently asking the latter whether it was through his grandmother or his grandfather that he was descended from a monkey. At this point Huxley, according to his own account of the meeting, whispered to his neighbor, the British surgeon Sir Benjamin Brodie, "The Lord hath delivered him into my hands."

Huxley waited patiently until Wilberforce was finished. When the audience called for him, he arose and soberly defended Darwin's theory as a legitimate scientific hypothesis. He pointed out that Darwin had implied not that human beings were descended from apes but that humans and apes had a common ancestor thousands of generations ago. And then Huxley delivered what was intended to be the fatal blow. This is his account of what followed:

> I asserted—and I repeat—that a man has no reason to be ashamed of having an ape for his grandfather. If there were an ancestor whom I should feel shame in recalling, it would rather be a man, who, not content with an equivocal success in his own sphere of activity, plunges into scientific questions with which he had no real acquaintance, only to obscure them by an aimless rhetoric, and distract the attention of his hearers from the real point at issue by eloquent digressions and skilled appeals to religious prejudice.

Supposedly, Huxley's response led to an uproar. According to stories told later, one lady fainted and had to be carried out. Spectators leapt up from their seats and shouted. Other speakers followed, each making his points in favor of or against evolution with vehemence. Finally, the last one to speak, the English botanist Joseph Hooker castigated Wilberforce, saying that he could not have read Darwin's book and that he could not be familiar with the basic ideas of the science of botany. Hooker then described his own conversion to belief in evolution. Wilberforce did not reply, and the meeting ended.

At least this is the story that has been repeated again and again in textbooks, in biographies, and in histories of evolutionary thought. But when we read it we should bear in mind that history is generally written by the victors. In this case, it was the evolutionists who were eventually victorious. One cannot help but wonder how the story would have been told if Darwin's theory had somehow been discredited.

As Stephen Jay Gould has pointed out, a somewhat different picture emerges when this now-mythical account of the meeting is compared to known facts. The evolutionists may not have been as triumphant as they claimed to be. For one thing, Huxley's words apparently did not carry very well, and they may have been inaudible to much of the audience. After the meeting, when Hooker reported the events to Darwin, he implied that this was the reason he had risen to Huxley's defense at the end. There is no contemporary account of Huxley's response to Wilberforce. The words I quoted are taken from one of his letters and may have embellished after the fact. It is true that a lady fainted. But most likely she fainted simply because it was hot, not because Huxley had replied to a bishop in a scandalous manner. Tightly corseted women commonly fainted in those days for any number of reasons, one of which was heat.

Contemporary reports of the meeting indicate that many people went away believing that Wilberforce had advanced the stronger arguments. But of course this skirmish had little effect on the course of events. Darwin's theory quickly gained acceptance. By the mid-1860s, the majority of British biologists were evolutionists, and the theory was widely discussed by members of the general public. To be sure, there were some who pointed out that belief in evolution was not consistent with a literal interpretation of Genesis and that the theory was therefore unacceptable on religious grounds. However, evolution quickly became a respectable, if not universally accepted scientific theory.

Darwin had numerous supporters. But there was one individual who supported Darwin's idea so vociferously that he became known as "Darwin's bulldog." This was Huxley. Huxley wrote about evolution in the leading periodicals of the day and defended Darwin against his scientific enemies, replying to criticisms and unfavorable reviews of *On the Origin of Species.* Huxley may or may not have triumphed over Wilberforce in the manner told. Nevertheless, he did more than any other individual, including Darwin himself, to gain acceptance for the idea that every living species was the product of evolution.

And yet Huxley did not agree with some of the major tenets of Darwin's theory. In *Origin,* Darwin had expounded his theory of natural selection at great length. According to Darwin, natural selection (I will explain this idea in detail shortly) was the major cause of evolution. Furthermore, Darwin said, evolution was a gradual process. Species evolved by the accumulation of small changes over long periods of time. The evolution of one species into another was a process that might require hundreds of thousands or even millions of years.

Huxley, on the other hand, doubted that natural selection was the only important factor in evolution. For example, in an essay written in 1878 Huxley stated, "It is quite conceivable that every species tends to produce varieties of a limited number and kind, and that the effect of natural selection is to favour the development of some of these, while it opposes the development of others along their predetermined lines of modification." This idea of "predetermined lines of modification" is one that was not to be found in Darwin's writings. In fact it was antithetical to his conception that it was natural selection that determined what the lines of modification would be.

Huxley was just as doubtful about Darwin's idea of gradualism. The day before *Origin* was published, he wrote to Darwin, "You have loaded yourself with an unnecessary difficulty in adopting *Natura non facit saltum* [Nature makes no jumps] so unreservedly." Huxley, who had been made privy to Darwin's ideas before the book was published, wanted to retain the idea that a species could give rise to a new species by a sudden leap, or saltation. He continued to believe this in spite of the arguments that Darwin made for gradualism in *Origin.*

Huxley was not alone in disagreeing with Darwin. In fact, most of Darwin's supporters in the period immediately following the publication of *Origin* took issue with him in one way or another. For example,

Hooker, who was as strong a supporter of Darwin as Huxley, also believed that Darwin attached too great an importance to natural selection. And the evolutionist British philosopher Herbert Spencer, who coined the term "survival of the fittest" (a term that was not used by Darwin in early editions of his book), did not believe in natural selection at all.

This raises the question "How could these individuals have supported Darwin if they did not believe in some of his most basic ideas?" The answer is not hard to find. In the years following the publication of *Origin,* belief in evolution as a natural process was sufficient to make one an "evolutionist." Few thought it was necessary to agree with Darwin on all points if one was to support him. It was realized even then that Darwin's "theory of evolution" was not a single idea but rather a bundle of a number of different, related, hypotheses.

Evolutionary Subtheories

According to the German-American biologist Ernst Mayr, Darwin's theory of evolution can be divided into five distinct subtheories, which are, for the most part, independent of one another. As Mayr describes them, Darwin's five theories (I am paraphrasing him here) are as follows.

1. *Evolution as such.* This is the idea that evolution takes place.

2. *Common descent.* This is the idea that every group or organisms (mammals, for example) is descended from a common ancestor, and that all organisms can be traced back to a single origin of life.

3. *Multiplication of species.* This is the idea that species multiply. They may do this by splitting into two distinct species at various different times during their evolution.

4. *Gradualism.* This is the idea that evolution is an accumulation of small changes. New types do not suddenly appear. That is, there is no saltation.

5. *Natural selection.* Evolution comes about because there is an abundance of genetic variation in every generation. Relatively few individuals survive and pass along their favorable genetic characteristics to the next generation.

These five ideas are not entirely unrelated. For example, if species do multiply, then it seems natural to assume that they must have had a common ancestor in the distant past. So theories 2 and 3 are not entirely independent of one another. Thus it would be possible to break Darwin's theory down into four subtheories instead of five. On the other hand, it is possible to break Mayr's five subtheories down further. Some authors have cited eight or more components. There is no particular reason to favor any of these groupings over any of the others. Determinations of the precise number of subtheories contained in Darwin's conception of evolution are somewhat arbitrary. However, it should be obvious that the idea that Darwin's theory is actually a bundle of different ideas is a valid one. Once this point is understood, it is easy to see how scientists such as Huxley could have counted themselves among Darwin's supporters when they disagreed with him on major points.

The breakdown of Darwin's theory into subtheories also enables us to understand how there can be controversies about the precise nature of evolution today, when all evolutionary biologists are in agreement on certain major points. It will therefore be useful to look at Mayr's subtheories in some detail. It might be best, however, if I change Mayr's order somewhat. For example, I think it would be most reasonable to consider natural selection first. After all, Darwin's theory was one of the evolution of species "by means of natural selection."

I don't plan to discuss evolution as such in any detail here, having covered it earlier. However, the various subtheories are interlinked, so you shouldn't be surprised if some of the material that follows constitutes additional evidence for the idea that evolution has indeed taken place.

Natural Selection

Natural selection is really a very simple idea. It is based on the observation that, in every species, there is always a great deal of variation. If two individuals are selected at random, they are likely to be different in a large number of different ways. Human beings, for example, are not all the same height. Nor are they all equally intelligent or equally strong. Some have better vision than others, and some are more susceptible to various kinds of diseases. We differ from one another in innumerable ways.

Darwin's idea had another important component. He realized that the individuals that constitute a species generally have many more offspring than can possibly survive. Generally, only a few live long enough to mate and produce offspring of their own. This has certainly been true of the human species for most of its evolutionary history. Until fairly recent times, large families were common. But the infant and child mortality rates were so high that only a minority survived to adulthood and had children of their own. In other species the disproportion between the number of offspring produced and the number that survive is even greater. For example, an average-size female Atlantic cod lays 2 million eggs in a breeding season, but on the average only two of them hatch, grow into adult fish, and survive long enough to successfully reproduce. As Darwin himself noted, even among slow-breeding animals such as elephants, numerous individuals die before reaching reproductive age. Darwin calculated that if this were not the case, the descendants of a single pair of elephants would number nearly 19 million after a period of only 740 or 750 years.

Organisms must therefore compete with one another for survival and reproductive opportunities. And, since these organisms are not perfectly alike, some of them are better suited for this competition than others. Those with the best hereditary endowment are the ones most likely to pass their genes along to the next generation. Even if certain genes confer only a very slight advantage, they gradually become more common over a long period of time. For example, if an antelope has hereditary qualities that allow it to run just slightly faster when it is trying to escape from a predator, these qualities will be likely to proliferate in future generations. The slower antelopes, who do not possess these qualities, will gradually die out. When this happens, the species as a whole will have evolved to be better adapted to its environment, an environment in which predators are present.

Natural selection causes species to adapt to specific environments. No matter how well adapted a land mammal is, it cannot survive in the depths of the ocean. And the adaptation of the camel to its environment does not enable it to survive in the Arctic. Naturally species can adapt to an environment in many different ways. Carnivores are animals that became specialized for capturing prey or for scavenging. Herbivores adapt to the types of foliage that are available in their surroundings. Parasites adapt to the hosts that are available. And of course even bacteria adapt. Pathogenic bacteria adapt to certain kinds of animal hosts, or by developing resistance to antibiotics. And of course the beneficial *E. coli* bacteria that help us digest our food (only certain strains of *E. coli* cause disease) have adapted to life in our intestines.

The organisms that populate Earth have been evolving for billions of years. Thus it is not surprising that some very complex adaptations have arisen. In some cases, they seem quite bizarre. For example, there is the brainworm, a parasite that infects sheep. It does not burrow into the sheep's brain; it received its name for quite a different reason. The brainworm has a complex lifestyle. Some of the worms are expelled in the sheep's feces. When the feces are eaten by snails, the worms take up residence in their new host. The snails then expel worm larvae in a mucus that is eaten by ants. Once an ant is infected, most of the worms remain it its abdomen, but one will make its way to the ant's brain and take up residence there. This causes the ant to exhibit a behavior that it would not ordinarily engage in. It crawls up a stem of grass, and remains there, patiently waiting to be eaten by a sheep.

Could this adaptation have arisen by natural selection? Of course it could. It would be impossible to work out all the details of the manner in which this behavior evolved. After all, we cannot travel back in time to observe what actually happened. Reconstructing the evolution of a pattern of behavior is much more difficult than gaining an understanding of physical evolution. Behavior, after all, generally leaves no fossils. However, the life cycle of the brainworm begins to seem a little less bizarre when one considers the fact that many different kinds of parasites spend part of their life cycles in snails. This was must have been the case with the evolutionary ancestors of the brainworm as well. Not many of the parasites that later wound up in ants would at first have gotten back to their original host, the sheep. That would have happened only from time to time by accident. However, if some small genetic change in the brainworm caused them to behave in such a way as to cause the ants to be just a little more likely to be eaten by sheep, natural selection could easily have improved the adaptation, producing the striking pattern that is observed today.

It is even easier to see how such a complex organ as the eye could have evolved. This was a question that troubled Darwin somewhat.

However, the pattern is reasonably clear today, because we can observe all the stages in the evolution of complex eyes in different kinds of living animals. For example, some single-celled animals have tiny light-sensitive spots on their bodies. This gives them some ability to move toward, or away from, a source of light. Some many-celled animals, such as certain flatworms and shellfish, have light-sensitive cells arranged in a little cup, called an eyespot. In other animals, such as the swimming shellfish *Nautilus,* the light-sensitive cup is deeper, and the opening that allows light into it is narrow, forming a primitive kind of lensless "eye" that functions something like a pinhole camera.

Once this kind of "eye" has evolved, the evolutionary steps that lead to eyes like our own are not particularly complex. The light-sensitive pit becomes covered with a transparent or translucent layer of skin. In this case the eye cavity becomes filled with cellular fluid rather than air or water. Part of the region that contains the cellular fluid evolves into a lens. Finally, full, complex eyes evolve. All these stages, and some intermediate ones, are observed in various different species of mollusks, while the octopus and squid have highly developed eyes that are similar to our own.

It is estimated that the complete evolution of a complex adaptation like a fully functioning eye requires about 2,000 steps over something of the order of 400,000 generations. In an evolutionary line with a generation time of one year, a complex eye could evolve in less than a half million years. If this is the case, the eye could have evolved more than once. Indeed, there is convincing evidence that it has. The eyes of the octopus and squid certainly evolved independently of our own, and it is thought that the eye evolved independently in a number of different evolutionary lines.

Natural selection cannot produce any kind of adaptation. It could not, for example, create cows that were able to fly. However, any genetic change that enhances the likelihood that individuals will survive and reproduce will be passed on from generation to generation.* Over a very great number of generations, the accumulation of small changes can have a dramatic effect. This is exactly what is observed in the case of the evolution of the eye. In many species, having light-sensitive eyespots is advantageous. If a slightly more complex system increases this advantage, then natural selection will favor greater complexity. This process can continue until very complex eyes like those of human beings evolve.

The Problem of Heredity

Darwin realized that there was one problem with his theory of natural selection. The theory depended on the transmission of hereditary characteristics from one generation to the next. But in Darwin's day, no one really knew how heredity worked. According to ideas that were prevalent at the time, inheritance was a "blending" process that combined the parents' characteristics. It wasn't difficult to observe that, if a tall man married a short woman, their children were likely to be approximately average in height. Matings between people of different races produced offspring whose skin was intermediate in color. If a very fast stallion was mated with an average mare, the offspring would not be likely to be able to gallop as rapidly as their sire.

Under such conditions, natural selection would simply not have worked. Any favorable characteristics that arose in a population would be washed out. For example, if inheritance were a blending process, an ante-

---

*Provided, of course, that a population is not too small. In a very small population, chance may cause favorable traits to be lost. For example, if three individuals out of a population of sixty possess some trait that enhances survival, they may all be killed by disease or predators before they are able to reproduce. On the other hand, if 300 individuals out of a population of 6,000 possess such a trait, extinction is very unlikely.

lope that was slightly faster than the others could not pass on this characteristic; its offspring would be indistinguishably faster than average.

Darwin never found a solution to this problem. In his day, nothing was known of genes or genetics. Today we understand that a favorable genetic mutation is likely to be passed on to succeeding generations, and that natural selection will cause it to spread throughout an entire population. The individuals possessing this gene, after all, are more likely to survive and to reproduce. Similarly, unfavorable genes or combinations of genes will eventually disappear. The individuals that possess them will be less likely to produce offspring. But Darwin could not have known this. Modern genetics was developed only after his death.

The first discoveries concerning genetic inheritance were made even before Darwin published *Origin*. They were made by an Austrian monk named Gregor Mendel, who had no reputation as a scientist and who published his results in an obscure scientific journal. Mendel's work remained unknown until it was rediscovered in 1900, some eighteen years after Darwin's death.

Mendel began his experiments in a monastery garden in 1856. He had obtained pea seeds of a number of different varieties but of the same species, so the varieties could easily be crossed. However, they exhibited different traits. For example, some produced white flowers, while others had purple flowers. Some of the pea plants produced yellow seeds, while others had seeds that were green. There were plants with wrinkled seeds, and plants with smooth round ones. In all, the pea plants varied in seven different ways.

Mendel crossbred his plants in numerous different ways to see how the different traits were transmitted from generation to generation. As he experimented, Mendel discovered that inheritance was not a matter of blending. He found, for example, that there was some factor in the makeup of purple-flowered plants that caused this trait to appear in their offspring.

The white-flowered plants contained a different factor that passed along the white color. When the two varieties were crossed, there was some blending. But the pure traits appeared again in subsequent generations. This showed that there were hereditary factors that did not disappear and that the commonly held views about inheritance were incorrect. Mendel didn't call these hereditary factors genes, however. That term was coined by the Danish biologist Wilhelm Johannsen in 1909.

Mendel's discovery was one of the most important in the history of biology, but it attracted no attention at the time. When he read a paper about his work at a meeting of the local society of natural history, there were no questions and no discussion. Mendel published his results in an obscure journal, *Transactions of the Brünn Natural History Society,* in 1865. But Mendel's work attracted no notice, even though the journal in which he published reached the major libraries of America and Europe, and he did little further scientific work after being made abbot of his monastery in 1868. When he died in 1884, his accomplishments were still unknown.

It wasn't until 1900 that Mendel's work was independently discovered by three European botanists, Hugo de Vries, Carl Erich Correns, and Erich Tschermark von Seysenberg. These three scientists had separately worked out the laws of inheritance, and when they searched the scientific literature on the subject, they discovered Mendel's papers. To their credit, each of the three gave priority to Mendel and presented their own work only as confirmation of his discoveries. The science of genetics had been born.

Genetics vs. Gradualism

As the science of genetics began to be developed during the early years of the twentieth century, the scientists who worked in the field tended to believe that the new results contradicted one of Darwin's other theories, that of gradualism. Many of them went so far as to conclude that the theory of natural selection had been superseded. Evolution, they concluded, could easily be explained in terms of genetic inheritance and mutation. Darwin had spoken of gradual change over long periods of time. But the first evidence discovered by geneticists seemed to indicate that the presence of different genes produced very noticeable effects. For example, Mendel's peas had either white or purple flowers. There was no continuous variation in between. And if the substitution of one gene for another had large effects, this contradicted Darwin's idea of gradual change.

Thus, during the early years of the twentieth century, scientists had differing ideas about the cause of evolutionary change. For the most part, naturalists and paleontologists saw evolution as the result of natural selection, while the geneticists emphasized the role of mutation. By the 1920s, however, it had been established that most mutations had small effects and were unlike the large mutations envisioned by the early geneticists. The accumulation of small mutations acted on by natural selection could cause evolutionary change, and combinations of genes could produce continuous variation, such as height in human beings.

During the 1930s, the British geneticist J. B. S. Haldane, the American geneticist Sewall Wright, and the British statistician Ronald Fisher succeeded in reconciling genetics with Darwin's theory. They mathematically analyzed mutation rates, rate of reproduction, and other factors, and showed that genetics and the theory of natural selection were not inconsistent. On the contrary, genetics could be used to strengthen Darwin's theory by providing it with a mathematical foundation.

Finally, during the 1930s and 1940s, it was shown that the findings of modern genetics were perfectly consistent with the patterns of long-term evolutionary change that had been discovered by the paleontologists.

Geneticists and biologists had brought their ideas together in what is sometimes called the "neo-Darwinian synthesis." Several decades previously, the significance of natural selection had been disputed. But after the synthesis was accomplished, Darwin's theory seemed more firmly established than ever.

Furthermore, it now seemed possible to conclusively refute certain ideas that had been common in Darwin's day and that were still believed by a minority of scientists. The sudden emergence of new forms—the saltation that Huxley and some of his contemporaries believed in—did not seem to be possible. Second, the inheritance of acquired characteristics was impossible. Environmentally induced changes in an organism had no effects on the genes. Building strong muscles by lifting weights would not cause one to have stronger children. The giraffe did not have a long neck because previous generations had lengthened theirs by stretching. And finally, there was no built-in drive toward evolutionary "progress" that caused species to evolve along certain lines. Evolutionary change came about when natural selection altered the frequency of certain genes in a population.

You shouldn't get the impression that the neo-Darwinian synthesis is a final theory that must be accepted without reservation and that cannot be modified. Religious dogma may have that character, but scientific theories do not. Science progresses because scientists commonly question generally accepted premises and test them by new, ever more accurate methods of observation and experimentation. If scientific ideas ever became dogmatic, then no new discoveries could be made.

Thus the conclusions of the scientists who created the neo-Darwinian synthesis should not be regarded as inscribed in unalterable form on some stone tablet. From time to time, modifications have to be made in every theory. However, the synthesis was an extremely successful theory. It created a single framework in which different kinds of knowledge about the organisms that inhabit Earth and their evolution could be understood. In fact, it was so successful a theory that today we find it extremely difficult to think of genetics as something separate from the rest of biology. It would be just as difficult to view evolution in anything but a genetic framework. We now know that spontaneous mutations take place and that the accumulation of small mutations over a long period of time must inevitably produce evolutionary change.

Molecular Biology

The next great step forward was the discovery of the structure of DNA by the American geneticist James Watson and the British biophysicist Francis Crick in 1953. Watson and Crick not only showed that DNA was a double helix composed of two long intertwined molecules, they also showed how DNA could duplicate itself. If the two strands of DNA were unraveled, a new strand could be formed along the length of either of the original ones. In the years that followed, this led to an understanding of how genes—which were segments of a long DNA strand—could duplicate themselves. It also led to an understanding of the causes of genetic mutation.

All cells—bacterial cells and eukaryotic cells alike—make use of a genetic code that is stored in long strands of deoxyribonucleic acid, which is commonly known as DNA. Each strand of the DNA double helix is constructed of four different molecules called nucleotides and chemicals that bind the nucleotides together. The nucleotides have been given the names adenine, guanine, cytosine, and thymine, which are generally abbreviated with the capital letters A, G, C, and T.

Every cell manufactures proteins. A cell must do this if it is to survive and reproduce; if the protein production ever ceased, the cell would quickly die. The manufacture of proteins is accomplished in the following manner: First, an enzyme (an enzyme is a special-purpose protein) separates two strands of a section of the DNA helix. Then, a strand of RNA (ribonucleic acid) is formed on the DNA template. RNA is also made up of nucleotides. These are A, G, C, and U. In RNA, U (uracil) is substituted for the T in DNA. This process transfers the genetic information in DNA to RNA. The cell then uses RNA to make proteins. Since the RNA is single-stranded, it does not have to be split apart by enzymes before this manufacture can take place.

The genetic information in DNA is encoded in sets of three nucleotides called triplets. Since there are four different nucleotides, there are some sixty-four possible combinations in all. The first nucleotide can be either A, G, C, or T. Four choices are also possible for the second and third. There are sixty-four possible combinations in all; $4 \times 4 \times 4 = 64$. Most of the triplets code for the creation of a specific amino acid. Three of them function as stop signs, marking the point at which the reading of the DNA is to stop. Thus a certain segment of DNA always gives rise to the production of the same protein. Proteins are nothing other than long chains of amino acids, and the stop signs tell the cell when the production of the protein is complete. The stop signs tell the cell that no more amino acids are to be added.

There is a certain amount of redundancy in this arrangement. Although there are some sixty-four different triplets, living cells make use of only twenty amino acids. For example, the presence of the triplet GUC in an RNA strand tells the cell to add a molecule of the amino acid called valine to the protein being manufactured. GUU, GUA, and GUG cause it to do exactly the same thing. But note that the redundancy cannot be eliminated by using doublets instead of triplets. There are only sixteen possible doublet combinations, and this is not enough to code for twenty amino acids.

The discovery of the structure and function of DNA suggested to scientists that it might be possible to study the genetic makeup of a cell. If some technique for doing this were discovered, it would be possible to analyze the DNA of both single-celled and multicellular organisms. Although the cells of a multicellular organism can take many forms—a liver cell does not look much like a skin cell, for example, and neither does it resemble the neurons that are found in the brain—all contain exactly the same DNA. Cells develop in different ways because they respond to chemical signals that they receive from other cells, but they are all genetically identical.

At first, it was cellular proteins that were studied. The DNA of a cell exists only in small quantities. On the other hand, numerous different copies of any given protein will be manufactured. In 1966, the Harvard geneticist Richard Lewontin (who was then at the University of Chicago) and his Chicago colleague J. L. Hubby refined a technique called electrophoresis, which had been discovered by the Swedish chemist Arne Tiselius around 1930, and used it examine some eighteen proteins that were found in the cells of the fruit fly *Drosophila*. Lewontin and Hubby examined the proteins of some five different populations of *Drosophila* and found that they could use the technique to obtain an estimate of the genetic differences between these populations. Meanwhile, the British geneticist Henry Harris did a similar study of ten proteins found in different populations. The results of both studies were published in the same year. Almost at once, many other geneticists began to use the technique, which was a relatively easy one and which could be used to study the proteins in the cells of any species.

Electrophoresis is a technique based on the fact that many proteins carry an electrical charge. Some are positively charged and some have a negative charge. If the proteins are placed between two electrodes, the positively charged proteins move toward the negative pole and the negatively charged proteins move in the opposite direction. If the proteins are dissolved in a gel, they move slowly, and proteins of different size and charge move different distances in any given period of time. This makes it possible to separate the proteins from one another and to determine which ones are present.

The technique can also be used to study segments of DNA directly. And indeed some studies of DNA were performed in the 1970s and 1980s. However, scientists were hindered by the fact that DNA was present in much smaller quantities than the proteins, and the study of DNA was a difficult and laborious process. But then, in 1983, Kary Mullis, then at Cetus Corporation, developed a process called the polymerase chain reaction, or PCR, an accomplishment for which he was awarded the 1993 Nobel Prize in chemistry. PCR is a technique that allows as many as a billion copies of a single piece of DNA to be made. It is a fairly simple process that makes use of a natural enzyme called polymerase that replicates DNA. It is so simple a process that almost any molecular biologist might have thought of it at any time during the decade that preceded Mullis's discovery. However, no one did. Sometimes it is the simple things that elude us.

Nowadays, the genetic makeup of entire organisms such as *E. coli* bacteria has been worked out. However, scientists have mapped the genomes of only two multicellular organisms. These are much more complicated. Where a bacterium has at most a few thousand genes, human beings—to cite one example—are made up of cells that contain about 100,000 genes,* which are divided among twenty-three pairs of chromosomes. The human genome is complicated in other ways also, genes are often repeated a number of different times in a strand of DNA, and there are long sections of DNA that do not code for proteins at all.

---

*This is a very approximate figure. At present, estimates range from around 30,000 to 150,000.

These noncoding segments, called introns, seem to perform no function. In effect, they are not part of the operating system of the cell. Introns are not seen in bacteria. The entire human genome is quite large; there are in all approximately 3 billion nucleotides.

By September 2000, only one human chromosome, chromosome 22 had been mapped. Chromosome 22 is a small one, and there are still some gaps in the mapping of its genetic structure. However, some progress has been made on the mapping of all twenty-three chromosome pairs, and a "rough draft" has been completed. The mapping of the human genome will probably be finished sometime during the first few years of the twenty-first century.

Molecular Biology and Darwin's Theory

Molecular biology provides convincing evidence for two of Darwin's theories, the ones that Mayr called "common descent" and "multiplication of species." The fact that all terrestrial organisms use the same genetic code is evidence that they are descended from a common ancestor. There is no particular reason why DNA codes for proteins in precisely the way it does. For example, it is possible to imagine cells in which the triplet GUC would tell the cell to make not valine but one of the other nineteen amino acids. This tells us that all living organisms are related to one another. Furthermore, cells in widely different organisms manufacture the same kinds of proteins. Many of the proteins found in human cells are also made by E. coli bacteria. The only reasonable conclusion one can draw is that all organisms are descended from some ancient ancestor with a particular genetic code.

It does not follow that life arose on Earth only once. As I pointed out previously, Earth experienced numerous collisions with astronomical objects during its early history, many of which released many times more energy than the collision that led to the demise of the dinosaurs. Many of these early collisions produced enough energy to cause the entire planet to become molten, if it wasn't molten already. It is possible that life came into being a number of times, only to be destroyed before it had much of a chance to evolve.

However, life did finally maintain a foothold. The record of life that begins 3.5 billion years ago is continuous. No one knows exactly what the first living organisms were like. There are a number of theories about the manner in which life may have begun, but no one can travel 3.5 billion years back in time to confirm or refute them. It is thought that life began either with proteins or with RNA, not DNA. It has been shown that under certain circumstances, proteins and RNA are able to replicate themselves. DNA, on the other hand, cannot replicate without assistance. Protein enzymes are needed to split the double-stranded DNA apart before it is able to replicate itself or to produce the proteins required for the splitting. This means that protein enzymes must already exist before the DNA can become functional.

So the most likely scenario is this: some kind of chemical evolution preceded the creation of life. Numerous different organic chemicals existed on the surface of Earth. It has been found that many organic chemicals, including amino acids, exist in space. Quantities of these chemicals would have been carried to the surface of Earth on falling interplanetary dust particles. Others could have come from the tails of comets that brushed Earth's atmosphere. Amino acids and other organic chemicals could also have been created in chemical reactions that took place on Earth. The ingredients necessary for the creation of life were undoubtedly present when Earth was very young. Under the right conditions they would have combined to form even more complex molecules. Some of these molecules were able to replicate themselves. It has been shown that some kinds of RNA have the ability to do this, and self-replication has also been observed in peptides (small proteins). As these self-replicating chemical systems gained complexity, they more closely resembled living cells. Finally, complex systems evolved that had all the characteristics of life.

One defect in this picture is the fact that a lot of details remain to be filled in. However, scientific understanding of the behavior of organic chemicals is great enough that it is possible to say that there is nothing implausible about a process of chemical evolution that led to life. In fact, the fact that life arose so quickly after conditions became favorable indicates that the creation of life may be something that is inevitable in any suitable environment.

Even if we could go back in time and see what actually happened, it might not be possible to determine the exact moment when a system of replicating chemicals became "alive." However, once living organisms did exist, they would have proliferated rapidly. The first life form would not have been subject to disease, and it would not have been in danger from any kind of predator. The organic chemicals that surrounded it would probably have provided a rich food supply. It would have quickly inherited the earth and have produced the many trillions of descendants that populate Earth today.

It is impossible to prove that all life on Earth can be traced back to a single origin. However, when all the evidence is taken into account, this idea seems very plausible. Thus there is at least circumstantial evidence for the existence of a common ancestor.

Evolutionary Branching

If the DNA sequences that make up the human genome are only partly known, how can we possibly say that humans and chimpanzees are

98.5 percent genetically identical? In reality, this statement is not as mysterious as you might think. There are a half dozen ways that the genes in the DNA of two organisms can be compared. For example, the proteins produced by the cells of the two organisms can be matched up. If a number of proteins are identical, they must have been produced by genes that are the same, or at least very similar. Alternatively, a number of known DNA segments can be compared. If one finds that they are, say, 90 percent identical, then it can be assumed that 90 percent is a good estimate of the similarity of the entire genomes.

There is even a method of determining the similarity of two strands of DNA without knowing their precise makeup. This method, called DNA–DNA hybridization, makes use of the fact that DNA is double stranded. DNA is extracted from two individuals of different species and cut into fragments that are about 500 nucleotides long (recall that G, C, A, and T are called nucleotides). When the DNA is heated in water, the double helix unravels, producing a collection of single strands. Single-strand mixtures from two different species are then allowed to hybridize; when they encounter one another, they will form a double helix. But now one strand of the helix will consist of DNA from one species, and the second strand has some from the other species.

If the two strands are very similar, they will bond together very strongly. If they are dissimilar, the bonds between them will be weak. The strength of the bonds can be determined by heating the hybrid DNA in a water bath and observing how readily the double helices break up, or melt. Weak bonds will break at relatively low temperatures. If the bonds are strong, more heating will be required. The relationship between similarity and temperature is a simple one. A 1-degree difference in melting temperature is roughly equivalent to a DNA difference of 1 percent. For example, if the hybrid DNA melts at a temperature that is 3 degrees lower than the temperature at which the original DNA from either species would melt, it can be said that the genetic similarity between the two species is 97 percent.

The fact a number of different methods for comparing the genetic distance between two species can be used allows scientists to be confident that their methods are reasonably accurate. If two methods yield approximately the same result, it can be concluded that the true genetic distance has been measured and that unknown factors of one kind or another are not skewing the results.

Knowledge of the genetic distance between species can be used to make inferences about their evolutionary past. For example, are human beings more closely related to the chimpanzee or the gorilla? If their bodily forms are examined, it is impossible to tell. Humans are more like chimpanzees in some ways and more like gorillas in others. The genetic evidence, however, tells us that it is the chimpanzee that is more closely related to us, and furthermore, it tells us that humans, chimpanzees and gorilla are more closely related to one another than to the third of the great apes, the orangutan. Therefore the chimpanzee must have split off from the evolutionary line that led to *Homo sapiens* more recently than the gorilla, and that the orangutan must have split off longer ago than either.

It is also possible to estimate the points in evolutionary time that these splits occurred. Studies of the genetic similarity between humans and chimpanzees can be used to infer that they probably began to evolve from a common ancestor about 5 million years ago.

When the figure of 5 million years was first announced by the biochemists Allan Wilson and Vincent Sarich of the University of California at Berkeley in 1967, their results provoked a storm of controversy. At the time, it was believed that the split had taken place at least 15 million and perhaps as much as 30 million years ago. This estimate was based on fossil evidence of a primate called *Ramapithecus*. It was then believed that *Ramapithecus,* which lived from 15 million to 8 million years ago, was the first direct ancestor of *Homo sapiens*. The only known *Ramapithecus* fossils consisted of teeth and jaw fragments. However, it was thought that these exhibited both humanlike and apelike characteristics.

It wasn't until 1982 that the controversy was settled. In that year, a complete fossil specimen of another early primate, called *Sivapithecus,* was discovered. It was apparent that *Sivapithecus* and *Ramapithecus* were closely related. Furthermore, the two primates had lived at approximately the same time. But *Sivapithecus* was also related to the ancestors of the modern orangutan. This fact caused anthropologists to conclude that *Ramapithecus* was not a human ancestor but an ancestor of the orangutan instead.

Anthropologists now agree that the evolutionary lines leading to the chimpanzee and human beings split off from one another about 5 million years ago. In 1994, Ethiopian scientists discovered some very apelike fossils at Aramis in Ethiopia. These fossils were dated to an age of about 4.5 million years, and the new fossil species was named *Australopithecus ramidus*. Some anthropologists considered *Australopithecus ramidus* to be an ancestor of chimpanzees, not of *Homo sapiens*. However, the fact that it is apelike in some respects and humanlike in others supports the conclusion that it lived very close to the time when the human and chimpanzee lines split. When these conclusions were drawn, the findings of the scientists who had relied on genetic evidence concerning human ancestry was conclusively vindicated.

Considerations of space prevent me from discussing the numerous other findings of molecular genetics in any detail. However, I think this one example shows how useful studies that compare the DNA of different species can be. Such studies not only allow scientists to fill in some of the details of evolution, they also provide convincing confirmation of Darwin's theory that new species evolve when an evolutionary line splits.

For example, there are some 4,000 different species of mammal on Earth today. In spite of the fact that there are numerous similarities between them, it would be possible to imagine that mammals evolved from reptiles some 4,000 different times, if the evidence did not tell us otherwise. But of course this is exactly what the evidence tells us. In Darwin's day, the existence of homologies was evidence that not just all mammals but all tetrapods had a common ancestor. But today it is not necessary to rely on homologies alone. The evidence provided by molecular genetics makes Darwin's theory seem incontrovertible.

Is There Still Room for Argument?

At this point, you may find yourself asking, "If all five of Darwin's theories of evolution have been confirmed, how is it possible that people are still arguing about them?" Hence I must point out that there is indeed room for argument and explain why.

Many of the arguments that are currently going on relate to natural selection in one way or another. I think it is safe to say that it has been conclusively shown that natural selection is the major cause of evolution. But it does not follow that it is the sole cause. There could conceivably be other factors that influence evolution in one manner or another. Even Darwin admitted that this might be the case. Although the full title of his famous book was *On the Origin of Species by Means of Natural Selection,* Darwin was careful to point out, "I am convinced that natural selection has been the main but not the exclusive cause of modification." Today, there are some scientists who would agree with that conclusion. And there are others who maintain that natural selection can account for all evolutionary change.

And then there is the matter of gradualism. It is agreed that evolution is a gradual process. But how gradual? Does "gradual" mean that it may take 50,000 or 100,000 years before natural selection can have noticeable effects? Or can a new species evolve in some shorter period of time, such as 10,000 years, or possibly even less? Does evolution always proceed at the same rate? If it does not, what are the implications for the theory of natural selection?

Is it possible to determine how particular adaptations that are found in living organisms evolved? If we can, what are the implications for our understanding of the behavior of human beings and other kinds of organisms? Behavioral traits evolve too, after all. A cat behaves like a cat because its genetic makeup causes it to do so. Evolution gave cats sharp claws, and it also provided them with traits that make them efficient predators of small animals.

These are only some of the questions that are being asked. And it should not be surprising that different scientists sometimes draw different conclusions. The essence of science, after all is questioning, not the creation, of rigid theories that have hardened into dogma.

Darwinian
"Fundamentalism"

I n his book *The Selfish Gene,* which was published in 1976, Richard Dawkins presented a "gene's eye view" of evolution. According to Dawkins, natural selection could be attributed to "selfish genes," which acted in their own best interest. Genes attempted to propagate as many copies of themselves as possible. If they were successful, they could endure for many millions of years. The individual organisms that carried the genes would eventually die. But if the genes were passed on to succeeding generations, they could achieve near-immortality.

According to Dawkins, biological organisms were only "gigantic lumbering robots" that the genes manipulated for their own ends. Since it was in the genes' interest to propagate themselves, they caused the organisms to behave in such a way that the genes would be propagated. Furthermore, the genes were ruthlessly selfish. They competed with other genes—found in other organisms—for survival. They acted to further their own ends and cared nothing about the good of the species.

Dawkins went on to say that the genes manipulated the organisms that carried them in many different ways. If it was in the genes interest that prey animals have longer legs so that they would be more likely to escape from predators, then the genes would cause them to evolve longer legs. If it was in the genes' interest that the males of a species attempt to spread their sperm around as widely as possible, this is what the males would do. And if it suited their purposes, genes would even cause the members of a species to cooperate with one another. Although selfish genes naturally gave rise to selfish organisms, mutual cooperation was in the genes' interest if it enhanced their probability of survival.

It sounds as though Dawkins was engaging in a kind of extreme genetic determinism. But this was not really the case. As Dawkins himself pointed out, he was propounding nothing but orthodox Darwinism. He naturally had no intention of ascribing conscious motives to genes. That was metaphor. He was only looking at natural selection from the point of view of the gene rather than that of the organism.

Dawkins's novel way of looking at natural selection really altered little or nothing of Darwinian theory. Emphasizing the role of the genes did not alter the fact that it was individual organisms that lived or died and that reproduced or failed to reproduce. Placing an emphasis on genes did not change the manner in which natural selection was described. The best-adapted individuals—those with the best genes—were still more likely to produce offspring and to pass those genes on to succeeding generations.

The ideas that Dawkins expressed were really not very new. Most of them had previously been stated by the American ecologist George C. Williams. In *Adaptation and Natural Selection,* which was published in 1966, Williams stated that "the laws of physical science plus natural selection can furnish a complete explanation for any biological phenomenon." In other words, natural selection alone could account for everything in evolution. It was not necessary to invoke any other supporting principles. Natural selection caused some genes to propagate themselves while others died out. And this could explain everything. At the time Williams wrote this book, there was considerable discussion about the possibility of group selection, a hypothetical process in which some groups of organisms would produce numerous offspring while other groups died out. Since Williams wanted to emphasize the fact that natural selection was the sole significant driving force in evolution, he argued at length against the claims of the scientists who wanted to give group selection a role. There was simply no evidence for group selection, Williams said. And if it existed, it would operate much more slowly than natural selection. Selection acted only on the individual.

If Williams were writing today, he probably would not place so much emphasis on this particular point, since group selection is no longer much discussed. The importance of his book lies in his emphasis on the gene as the unit of selection. As we have seen, this is the view that Dawkins was later to emphasize. Today, the Williams–Dawkins viewpoint is widely accepted. One could even say that it has become orthodox.

In other words, evolutionary biologists have come to consider natural selection to be even more important than it was thought to be in Darwin's day. More often than not, they reject the idea that evolution depends partly on other mechanisms. For example, one idea that has long been discarded is the nineteenth-century notion of the existence of some innate drive toward evolutionary "progress." There seems to be no plausible way in which such a drive could operate, and it is not necessary to invoke such an idea to explain biological phenomena. For example, the human brain did not evolve to its present size because some internal force was driving it. The human brain became larger because individuals with slightly bigger brains were more likely to survive and to produce offspring. Williams's idea that only natural selection is significant has now become commonplace.

Natural selection acts on the individual or on the individual genes, if one wants to use Dawkins's way of looking at things. According to the orthodox view, evolution happens because some genes survive and become part of the makeup of large numbers of individuals, while other genes perish. Obviously, natural selection is not capable of doing anything else. Living organisms are subject to the same laws of chemistry and physics that govern the behavior of inanimate matter. Horses could not evolve wings and learn to fly, or at least they couldn't unless they evolved into something very much unlike modern horses first. Their weight and the aerodynamics of flight make flying horses an impossibility. Similarly, no cow or pig will evolve the ability to jump over the moon—or even a small mountain. Or, to give a more mundane example, elephants could not evolve relatively spindly legs like those of horses. Such legs could not bear an elephant's weight; they would break every time the animal took a step. And elephants with broken legs would not be very likely to reproduce. Natural selection causes organisms to become better adapted to their environments. But there are constraints. If it is indeed true, as such scientists as Williams and Dawkins claim, that natural selection is the only factor that is important in evolution, it does not follow that anything is possible.

When Dawkins placed a great emphasis on natural selection as the only significant force in evolution, he wasn't saying anything very new. Many of his contemporaries were saying the same thing. Dawkins's only really original contribution was his metaphor of the "selfish gene." He believed that using this metaphor would make it possible to describe the workings of natural selection in an especially clear manner. It was a way of looking at things that would clarify the mechanisms of evolutionary change.

But not all evolutionary biologists agree. For example, Stephen Jay Gould takes issue with Dawkins's account of natural selection, calling

Dawkins a "Darwinian fundamentalist." According to Gould, Dawkins places too much emphasis on natural selection as the sole causal factor in evolution. Natural selection, he says, does not "explain why the world houses more than 500 species of beetles and fewer than fifty species of priapulid worms." (Priapulids are predatory, marine, mud-inhabiting worms.) Natural selection, Gould says, doesn't explain why most of the DNA in multicellular organisms "does not code for any enzyme or protein involved in the construction of the organism." It doesn't explain, he says, why the dinosaurs died out while the mammals survived. Natural selection, he concludes, may explain individual adaptations, but it doesn't explain long-term evolutionary trends.

Gould does not deny that natural selection is the most important factor in evolution. "Yes, eyes are for seeing and feet are for moving," he says. "And, yes again, I know of no scientific mechanism other than natural selection with the proven power to build structures of such eminently workable design." But he maintains that there is more to evolution than that. He sees terrestrial evolution as a "messy" process that cannot be understood in terms of any single fundamental principle.

Gould's most important contribution to evolutionary thought is the theory of punctuated equilibrium, which he and the American paleontologist Niles Eldredge proposed in 1972. Punctuated equilibrium is discussed at length in the next chapter. But I will summarize the main ideas here in order to paint a clearer picture of the approach that Gould and his colleagues take to evolution. According to Gould and Eldredge, evolution is not the slow, steady process that many scientists imagine it to be. Species did not evolve by slowly accumulating genetic changes over periods of many millions of years. Evolution, Gould and Eldredge said, took place in short bursts when new species evolved. Once a species had been created, it generally experienced a long period of evolutionary stasis, remaining very much the same until it became extinct. Evolutionary change, according to Gould and Eldredge, was something that typically took place over periods of tens of thousand of years, not over millions of years, as had previously been thought.

The theory of punctuated equilibrium did not contradict the idea that evolution happened because natural selection brought about the gradual accumulation of useful variations. After all, evolution is still "gradual," compared to the human life span, whether it takes place in 50,000 years or a million. However, the theory did set Gould and Eldredge to thinking about the roles played by speciation (the evolution of new species) and of extinction in the evolutionary process. According to Gould, some evolutionary lineages give rise to many more species than others, and the extinction rates may vary between lineages, too. If this is true, then some lineages produce more offspring than others, just as some individuals reproduce more prolifically. The beetle lineage, for example, seems to be an especially successful one, while the priapulid worms are not. This led Gould and Eldredge to believe that there existed large-scale patterns in evolution that did not depend on natural selection alone.

According to Gould, catastrophic extinctions, such as the one that led to the demise of the dinosaurs some 65 million years ago, also play a role in molding long-term evolutionary patterns. This extinction, which is believed to have been caused by the collision of Earth with an asteroid, did not affect just the dinosaurs. On land, mammals survived, and plants were relatively little affected. However, there were massive extinctions of the animals that inhabited the oceans. All in all, over half of the living aquatic species became extinct.

This extinction was not, however, the greatest Earth has experienced. In the Permian extinction, which took place about 245 million years ago, 95 or 96 percent of all living species perished. The Permian extinction, by the way, was not caused by a collision with an extraterrestrial body. Or at least there is no evidence to that effect. It seems to have come about when all Earth's continents drifted together to form a single supercontinent, called Pangea. It is not certain that the coalescence of the continents was the only cause of this extinction. But it must certainly have contributed to it. There were certainly dramatic climatic changes at this time. The fact that, as the continents came together, there was much less coastline certainly had an effect, as well. At the time, most organisms lived in shallow water. When the continents came together, there were fewer places they could live, and the range of environments that were available to them was undoubtedly smaller.

There have been numerous other mass extinctions in the history of Earth. Most likely there were a number of different causes. Collisions caused some extinctions but not all of them. And in fact, some extraterrestrial collisions appear to have had relatively little effect on life. For example, two large bodies collided with Earth about 36 million years ago. But the next major extinction event happened some 2 million years later. It appears that there was no causal connection between the collisions and the later extinctions.

Gould is fond of pointing out that, if an extinction had not wiped out the dinosaurs 65 million years ago, then we would never have evolved. He also observes that if the Permian extinction had not taken place, then the history of life on earth would be something very different. Though all this is true, it is not something that evolutionary theorists like Dawkins would deny. They agree that evolution has no "goal," and they are perfectly comfortable with the idea that evolution might have followed some other path. The belief that natural selection is all-powerful is perfectly compatible with the idea that human beings, and indeed intelligent life, could never have evolved under other circumstances.

   In one sense orthodox Darwinians like Dawkins agree with what Gould says about extinctions and about contingency—the idea that evolution could have followed an entirely different course under different conditions. Nevertheless they disagree about the significance of this fact. The orthodox Darwinians look at evolution in reductionist terms, while Gould looks for the existence of broad complex patterns. Even when they agree about the facts, they view the history of life on Earth from different perspectives.

   When I say that the orthodox Darwinists are "reductionist," I am not using that word in any pejorative sense. A reductionist approach has been characteristic of scientific thought since the time of Galileo, and it is reductionist methods that have been largely responsible for the phenomenal success of physical science. Physicists have successfully explained the nature of matter by identifying the fundamental subatomic particles and explaining their interactions with one another. Molecular biologists have explained the workings of genes by probing the structure and gaining an understanding of the functions of DNA. And geneticists have provided us with an understanding of natural selection by doing mathematical analyses of gene frequencies. Science is largely reductionist because, if it were not, we would have little science at all. If orthodox Darwinists seek to tie all evolutionary phenomena to the working of natural selection, they are only following a time-tested scientific tradition.

   But there are some cases in which pure reductionist methods are not adequate. For example, an understanding of the properties of water or air molecules does not explain the nature of turbulence. Turbulence can be understood only if a system (flowing air or water) is considered as a whole. Many of the properties of water and air can be explained in terms of their component molecules, but other properties—such as turbulence—appear at higher levels of complexity.

I return to a discussion of this later. I bring it up here only because I want to emphasize the point that sometimes Gould (and his colleagues) and Dawkins (and his colleagues) do not disagree about facts. Sometimes they disagree because they view evolution from different perspectives. Dawkins is very close to being a pure reductionist. Gould, on the other hand, looks for complex patterns in nature. They sometimes disagree because they look at evolution from different viewpoints and because each apparently thinks his viewpoint is superior to the other's.

Spandrels

The differences between Gould and Dawkins are not confined to difference in outlook, however. Some very specific criticisms of the practices of orthodox Darwinians can be found in a paper published by Gould and Richard Lewontin in 1979. Titled "The Spandrels of San Marco and the Panglossian Paradigm: A Critique of the Adaptationist programme," the paper began with a description of the central dome of St. Mark's Church (*San Marco* in Italian) in Venice. The dome is supported by two arches that meet at right angles. The arches divide the dome into four tapering triangular spaces. As Gould and Lewontin point out, these spaces are a byproduct of mounting a dome on two rounded arches; the arches could not divide the inner surface of the dome in any other way.

In these spaces—called spandrels—are mosaics of the four biblical evangelists and mosaic images representing the Tigris, Euphrates, Nile, and Indus rivers. Gould and Lewontin pointed out that the spandrels were not created for any architectural purpose. On the contrary, they are "non-adaptive" side effects of the architecture of the church. If the dome was to be supported by two interesting arches, the spandrels had to be there. They weren't created for the purpose of housing mosaics; they were decorated because there were empty spaces to be filled.

According to Gould and Lewontin, something similar frequently happens in evolution. Organisms, they say, have many traits that were not molded by natural selection. The traits exist because they are a by-product of something else. This does not mean that these traits are not useful. Once a spandrel exists, natural selection may modify it in some useful way, just as the builders of San Marco found that the triangular spaces could be decorated.

Gould has cited the human ability to read and write as an example of a spandrel. Natural selection caused our brains to become big, he says, "for a small set of reasons having to do with what is good about brains on the African savannas." But once they became big, human brains were able to do numerous different things that had nothing to do with the reasons natural selection caused them to grow larger in the first place.

Gould and Yale University paleontologist Elisabeth Vrba have devised the term "exaptation" to illuminate the role played by spandrels. Exaptations are spandrels that organisms have adapted for some useful purpose. They were not initially developed by natural selection for their current role, so they are not the same as adaptations, which were. Thus the swiftness of an antelope, which presumably evolved because natural selection weeded out the slower individuals, is an adaptation, while the human ability to read and write would be better described as an exaptation.

Dr. Pangloss Returns

When Gould and Lewontin referred to the "Panglossian paradigm" in the title of their paper, they were alluding to the ideas espoused by Dr. Pangloss in Voltaire's novel *Candide*. In this book, Voltaire satirized the doctrines of the German philosopher Leibniz, who had maintained that this was the best of all possible worlds. According to Dr. Pangloss, in this best of all worlds, everything existed for a purpose. For example, in explaining to Candide why he had contracted syphilis, Dr. Pangloss said, "It is indispensable in this best of all possible worlds. For if Columbus, when visiting the West Indies, had not caught this disease, which poisons the source of generation, which frequently even hinders generation, and is clearly opposed to the great end of Nature, we should have neither chocolate nor cochineal."

According to Gould and Lewontin, many evolutionists were guilty of interpreting the structures seen in living organisms in a similar way. They tried to break organisms down into distinct "traits" and then attempted to explain these traits as "structures optimally designed by natural selection for their functions." If they were unable to accomplish this to the degree they wished, then they spoke of "trade-offs." If natural selection could not optimize each part without imposing expenses on the others, then it found the best possible compromise between the various competing demands. They were sure that they could do this because they believed that natural selection was all-powerful, the "primary cause of nearly all organic form, function and behavior."

But in reality, the Panglossian program was a mistaken one, Gould and Lewontin said. Not only were some of these "traits" biologists saw in organisms likely to be spandrels, it was not even certain that this kind of breakdown was a valid procedure. Organisms, after all were "integrated entities, not collections of discrete objects." Gould and Lewontin believed that many evolutionists were engaging in a kind of ultrareductionism that was not always valid.

Gould and Lewontin did not confine themselves to criticizing the procedure of breaking organisms down into "traits." They attempted to give a number of examples of cases where the adaptationist program did not work very well at all. They pointed out that chance alone could cause some genes to occur more frequently or less frequently in a population. In some cases, chance could even cause genes to become widespread while natural selection was attempting to weed them out. Second, natural selection operating on one trait could cause another trait to change for no adaptive reason. As an example, they cited the fact that increase in brain size seems to be correlated with increase in body size in the same manner in all major vertebrate groups. The fact that an organism tended to have a brain of a certain size was often the result of natural selection acting on the size of its body. If, over long periods of evolutionary time, the body grew larger, the brain would follow suit and become larger too, whether additional cranial capacity was needed or not.

Furthermore, Gould and Lewontin pointed out, there could be selection without adaptation and adaptation without selection. As an example of the latter, they pointed out that the differing forms taken by sponges and corals apparently depended upon the ways in which water flowed in the environments in which they lived, not on genetic differences. The corals and sponges were plastic enough to adjust themselves in different ways to environmental conditions in different geographical regions.

There were often cases, Gould and Lewontin went on, where species of related organisms, or subpopulations of a single species, adapted in different ways to the same or similar environmental conditions. Although natural selection was operating, it could sometimes lead to a variety of different adaptive solutions. Sometimes there was no single optimal design, and different pathways could be followed.

And finally, the two authors said, there were spandrels. Organisms could develop traits that were by-products of selection. To be sure, spandrels often turned out to be useful when adapted for some purpose.

However, Gould and Lewontin said, the spandrels originally evolved for secondary reasons. They could not be directly attributed to natural selection.

Counterattack

Dawkins initially responded relatively mildly to Gould's criticisms. For example, in his book *The Extended Phenotype,* Dawkins characterized Gould's and Lewontin's antiadaptationist arguments as "unfair." There were many kinds of adaptationists Dawkins said. A good Darwinist would proceed in a more rigorous manner than that which Gould and Lewontin described. The true Panglossians, Dawkins said, were those who believed in group selection or in the evolution of traits "for the good of the species." The latter, especially, were guilty of sloppy thinking. Natural selection acted only to produce well-adapted individuals, even in cases where the good of the individual was contrary to the "good of the species."

Here, Dawkins was again echoing Williams. Group selection was not a topic that Gould and Lewontin had brought up in their paper. However, Dawkins apparently felt that Gould's and Lewontin's criticisms had been misdirected, so he brought up the issue himself.

A few years later, Dawkins began to counter Gould's arguments (he was now focusing on Gould; although Lewontin had also been an author of the paper on spandrels, Gould had been the more vocal critic) more forcefully. In his contribution to *The Third Culture* (Simon & Schuster, 1995), a book edited by literary agent John Brockman, Dawkins argued that Gould's "pluralist" view of evolution was based on a misunderstanding. Gould just didn't see that there was a real distinction between "replicators" (genes) and "vehicles" (individuals, groups and species). There was only one replicator known to biological scientists, Dawkins maintained.

And that was the gene. The vehicles were not replicators, so natural selection could act at only one level.

Dawkins also charged that Gould seemed to be saying things that were more radical than they really were. "He pretends," Dawkins said. "He sets up windmills to tilt at which aren't serious targets at all." In other words, Gould's ideas were really not much of a challenge to orthodox theory. He blustered, but many of the things he said were devoid of significant content.

Following Dawkins, Daniel Dennett entered the debate. His response was considerably more vehement. Gould, Dennett said in his book *Darwin's Dangerous Idea* (Simon & Schuster, 1995), was a would-be revolutionary who had mounted a series of attacks on conventional Darwinism over the years. He had done this so often that he was seen in some quarters as the "Refuter of Orthodox Darwinism." But, in reality, none of these attacks had really amounted to much. They had proved to be "no more than a mild corrective to orthodoxy at best."

Here, Dennett was doing little more than repeating Dawkins's charges. But there was more to come. Unfortunately, Dennett went on, Gould was the best-known writer on evolution. His revolutionary pose had had an impact that had been "immense and distorting." Gould, Dennett claimed, was responsible for causing many nonbiologists to believe that a number of tenets of Darwinian thinking had been discredited. But of course, nothing of the sort was true. Though Gould had gone "from revolution to revolution," evolutionary theory had remained relatively unchanged.

Before I go any further, I think it would be a good idea to say something about the themes of Dennett's book to put his remarks in the proper context. According to Dennett, "Darwin's dangerous idea"—his theory of natural selection—was one that many people found threatening. They believed that if life on Earth, including human beings, really did evolve by natural selection, then it might be necessary to draw conclusions that many people wanted to avoid. If natural selection was a "mindless, purposeless, mechanical process," then there could be no foundation for morality. Furthermore, "the illusion of our own authorship, our own divine spark of creativity and understanding" would be dissolved. A common fear about Darwin's idea was that it would "not just explain but *explain away* [Dennett's italics] the Minds and Purposes and Meanings that we all hold dear."

The implications of Darwin's idea had been resisted from the very beginning, Dennett said, by those who were afraid of its implications. There had been numerous attempts to chip away at the theory of natural selection, and Stephen Jay Gould was among those who attempted to do so.

Dennett's description of natural selection was very much like Dawkins's. Dennett used a different kind of terminology, however. Rather than speak of "selfish genes," he described natural selection as an "algorithmic process."

An algorithm is a rule for solving a problem, usually a mathematical one. It is not necessary for the person who uses the algorithm to understand why it works. Provided that no mistakes are made, the algorithm will always provide the correct answer. The long division we learn in school is an algorithm. Similarly, there are algorithms for finding square roots. In high school algebra we are taught methods for solving quadratic equations. These are algorithms, too.

The problem solved by an algorithm does not have to be a mathematical one. For example, the rules for selecting the teams that will be in the National Football League play-off constitute an algorithm. So do the rules for giving a wine a numerical score in a wine tasting. An algorithm may include an element of chance. For example, at one time, the rules for selecting the challenger to the world chess champion specified that a coin be tossed to choose the "victor" if other methods of breaking a tie between two contenders failed. The player who won the toss would advance to the next round while his opponent would be eliminated.

But even if using an algorithm does not always yield the same result, it does guarantee that some result will be obtained if the rules are followed. It is not possible to use an algorithm and obtain no answer at all. It couldn't be. The purpose of algorithms is to prevent this.

According to Dennett, natural selection is a kind of algorithm, too. Selection acts on the frequencies with which particular genes exist in a population. It is a blind process that weeds maladaptive genes out and promotes the propagation of good ones. It is not the kind of algorithm that must always produce the same result. If we could somehow run evolution all over again, there would be no guarantee that the same kinds of organisms or even similar ones would evolve.

Dennett sums up his concept of natural selection as an algorithm in the following words:

> Here, then, is Darwin's dangerous idea: the algorithmic level *is* the level that best accounts for the speed of the antelope, the wing of the eagle, the shape of the orchid, the diversity of species, and all other occasions for wonder in the world of nature. It is hard to believe that something as mindless and mechanical as an algorithm could produce such wonderful things. No matter how impressive the products of an algorithm, the underlying process always consists of nothing but a set of individually mindless steps succeeding one another without the help of any intelligent supervision; they are "automatic" by definition: the workings of an automaton. They feed on each other, or on blind chance—coin flips if you like—and on nothing else.

Gould, Dennett said, had time after time tried to chip away at this view of evolution. "Gould's ultimate target," Dennett went on, "is Darwin's dangerous idea itself; he is opposed to the very idea that evolution is, in the end, just an algorithmic process." He went on to suggest that Gould must have some "hidden agenda."

In due course, Dennett speculated about what this hidden agenda might be and suggested that Gould might have political or religious reasons for opposing the algorithmic view of natural selection. "Gould has never made a secret of his politics," Dennett said. "He learned his Marxism from his father, he tells us, and until quite recently he was very vocal and active in left-wing politics." Perhaps some of his theoretical stances reflect his "Marxist antipathy for reform playing itself out in biology."

Or perhaps Gould's antipathy to Darwin's dangerous ideas was a result of his "religious yearnings." Gould often quoted the Bible in his writings, Dennett noted. "One way of interpreting Gould's campaigns within biology over the years," he said, "might be an attempt to restrict evolutionary theory to a properly modest task, creating a *cordon sanitaire* between it and religion." In other words, perhaps Gould wished to give God a wider role to play than was allowed for by an algorithmic view of natural selection.

Dennett did not confine himself to speculating about Gould's motives. He also attempted to criticize Gould's theories in detail. An entire chapter of Dennett's book is devoted to criticisms of Gould's ideas. For example, Dennett includes a long discussion of spandrels. He begins by discussing the architecture of spandrels and attempts to show that they are really not as good an example as Gould and Lewontin had claimed. Then he attempts to downplay Gould's and Lewontin's argument, noting: "The thesis that every property of every feature of everything in the living world is an adaptation is not a thesis that anybody has ever taken seriously." Furthermore, natural selection could still be the sole cause of evolution even if some features of organisms were not adaptations.

In other words, maybe some of the features of organisms are spandrels, but so what?

Dennett sums up this chapter of *Darwin's Dangerous Idea* by claiming that Gould's "self-styled revolutions" weren't so revolutionary after all. Gould had made some good points, and some of his ideas had been incorporated into evolutionary theory. Some of his other ideas had simply been mistaken and could be dismissed. In the end, Dennett claimed, Darwin's dangerous idea had emerged "strengthened, its dominion over every corner of biology more secure than ever."

Thrust and Counterthrust

Gould responded in two related essays that were published in the June 12 and June 26, 1997, issues of *The New York Review of Books*. He began the first essay, titled "Darwinian Fundamentalism," by noting that Darwin himself had not believed that natural selection was the sole cause of evolutionary change. Gould cited the often-quoted statement that Darwin had made in the introduction to *Origin of Species:* "I am convinced that natural selection has been the main but not the exclusive means of modification." Darwin's theory, Gould implied, had always been pluralistic.

Nevertheless, Gould said, a group of evolutionists who wished to "out-Darwin Darwin" by placing emphasis on one component of Darwin's thinking had come to prominence. This group included Richard Dawkins, John Maynard Smith, and Daniel Dennett. These "ultra-Darwinists" or "Darwinian fundamentalists" shared a conviction "that natural selection regulates everything of any importance in evolution, and that adaptation emerges as a universal result and ultimate test of selection's ubiquity." These evolutionists, Gould said, "push their line with an almost theological fervor."

Next, Gould defended himself against the charges that had been leveled against him by Dennett and other thinkers. "Since the ultras are fundamentalists at heart, "Gould said, "and since fundamentalists try to stigmatize their opponents as apostates from the one true way, may I state for the record that I . . . do not deny either the existence and central importance of adaptation, or the production of adaptation by natural selection."

But, Gould insisted once again, natural selection could not explain everything. There were other, secondary, causes of evolutionary change. And appealing to these causes was not, as Dennett and others had claimed, an attempt to "smuggle purpose back into biology." These additional principles, Gould said, "are as directionless, nonteleological, and materialistic as natural selection itself."

Gould continued by arguing his long-held conviction that other factors besides natural selection had to be considered if one wanted to explain the evolutionary history of life on Earth. And then he turned to a discussion of *Darwin's Dangerous Idea*. Dennett's book, Gould said, presented itself as "the ultras' philosophical manifesto of pure adaptationism." Dennett explained the adaptationist view well enough, Gould went on, but he defended "a miserly and blinkered picture of evolution." Dennett's "limited and superficial book reads like a caricature of a caricature," Gould charged. Dawkins had "trivialized Darwin's richness by adhering to the strictest form of adaptationist argument in a maximally reductionist mode." Dennett, as "Dawkins's publicist," had managed to "convert an already vitiated and improbable account into an even more simplistic and uncompromising doctrine. If T. H. Huxley had truly been "Darwin's bulldog," Gould said, then Dennett had to be thought of as "Dawkins's lapdog."

Gould concluded the article by lashing out at John Maynard Smith, who had made some critical comments about Gould while reviewing Dennett's book for the same publication. In this review, Maynard Smith had said:

> Gould occupies a rather curious position on his side of the Atlantic. Because of the excellence of his essays, he has come to be seen by non-biologists as the preeminent evolutionary theorist. In contrast, the evolutionary biologists with whom I have discussed his work tend to see him as a man whose ideas are so confused as to be hardly worth bothering with, but as one who should not be publicly criticized because he is at least on our side against the creationists.

Gould quoted this passage in his essay, commenting, "It seems futile to reply to an attack so empty of content, and based only on comments by anonymous critics; if they were named, they would, I suspect, turn out to be a very small circle of true believers."

Gould went on, "Instead of responding to Maynard Smith's attack against my integrity and scholarship, citing people unknown and with arguments unmentioned, merely remind him of the blatant inconsistency between his admirable past and lamentable present." Perhaps Maynard Smith had gotten caught up in "apocalyptic ultra-Darwinian fervor," Gould suggested, and added that he was "saddened that his once impressive critical abilities seem to have become submerged within the simplistic dogmatism epitomized by Darwin's dangerous idea.

The Pleasures of Pluralism

It must have seemed to many at this point that the debate had degenerated into name-calling on both sides and that little more could be said. However, as it turned out, there was more to come. Two weeks after Gould's first essay appeared, *The New York Review of Books* published a second, titled "Evolution: The Pleasures of Pluralism." In this essay, Gould returned to the subject of the content of Dennett's book, and he also criticized the methods of a relatively new scientific discipline called evolutionary psychology.

I devote a separate chapter to evolutionary psychology in which I discuss the claims of its proponents and the criticisms of scientists who are skeptical about them. Here I discuss only Gould's responses to the charges that Dennett made against him in *Darwin's Dangerous Idea.*

Gould observed that Dennett had devoted the longest chapter of his book to criticisms of Gould's ideas. In his essay, Gould responded by saying that the only real argument that he could find amid Dennett's "slurs and sneers" was the charge that he (Gould) had tried to criticize what Dennett took to be "the true Darwinian scripture" by claiming "revolutionary" status for some of his ideas. After outlining Dennett's criticisms in detail, Gould charged that Dennett's "critique of my work amounts to little more than sniping at false targets of his own construction" and that Dennett had proceeded by "hint, innuendo, false attribution, and error."

Furthermore, Gould said, he had never made any claims about the "revolutionary" significance of his ideas. On the contrary, he had studiously avoided doing that. "Dennett can't support his charge of revolution-mongering with any quotation from me," Gould said. For that matter, he said, Dennett did not even seem to have a firm grasp of the material he attempted to discuss; witness the number of errors of fact in Dennett's book.

Finally, Gould said, Dennett had engaged in gratuitous speculation about his (Gould's) motives. "Dennett has no clue about my political or religious views," Gould said. Nevertheless, Dennett had engaged in a "little red-baiting," and had commented on Gould's religious views.

According to Gould, Dennett had been trying to make him into a "closet theist" by making reference to the fact that Gould frequently quoted the Bible in his writings. However, Gould maintained, he had actually quoted the Bible as great literature, and one passage that Dennett had taken to be a Biblical quotation hadn't been at all. Gould had actually been quoting the words of a famous African-American folk song.

Carnage on the Battlefield

Some of the people who read Dennett's book and also Gould's essays in *The New York Review of Books* probably looked at the carnage on the battlefield and concluded that Dennett and Gould were engaged in a very personal controversy. If they did, they were badly mistaken. Dennett is not the only individual who considers natural selection to be the only important cause of evolution. And Gould is not the only scientist who argues for a more pluralistic interpretation. In fact Gould names Dennett, Dawkins, and Maynard Smith as the triumvirate of narrow selectionist thinking.* Meanwhile, scientists such as Niles Eldredge and Richard Lewontin generally align themselves with Gould.

The group that Gould calls the "fundamentalists" believe that all biological change is the result of natural selection acting on genes. The members of the opposing camp believe that natural selection is the most important cause of evolutionary change but not the only one.

Who is right? There is only one way a question like this can be answered, and that is by looking at the scientific evidence. The differences

---

*The name of George Williams should probably be added to these three. After all, it was Williams who originally developed many of the ideas that the orthodox Darwinians expound. However, he has not become involved in the ensuing controversies to the extent that Maynard Smith, Dawkins, and Dennett have.

between the two camps of which Gould and Dawkins are members are partly philosophical. The members of one group endorse a reductionist outlook while the members of the other say that reductionist methods cannot give a complete explanation of large-scale evolutionary patterns. But there are also points on which new evidence could be brought to bear. As I show later in the book, scientists are constantly making new discoveries about the details of evolution, and it is likely that some of these discoveries will eventually shed some light on the controversies I am describing.

However, it is not yet time to talk about these discoveries. I have not yet talked about Eldredge's and Gould's theory of punctuated equilibrium. I have not yet talked about the work that is being done in a relatively new scientific field called the "sciences of complexity," and I have, so far, made only a passing reference to evolutionary psychology.

Dennett and Gould may be the most vocal participants in the controversies that are currently going on. But they are not the only ones. The question of how rigidly Darwin's theory of natural selection should be adhered too is one that has been coming up again and again in recent years, in a number of different ways. No one (except perhaps those who practice the pseudoscience of creationism) entertains any thoughts of overthrowing Darwin's theory. But there are numerous scientists who believe that something should be added to the theory.

And that is what this book is really about.

How Gradual Is Evolution?

Darwin always empha-sized that evolution was a gradual process. He argued against the saltationists among his contemporaries by insisting that there were no sudden evolutionary leaps. On the contrary, he insisted, over millions of years, natural selection produced a series of tiny changes. In time, these changes accumulated and caused new kinds of organisms to evolve.

Darwin was well aware that the fossil record did not show the small gradations that one might expect. Paleontologists did not discover series of fossils showing gradual evolutionary change. Instead, a fossil species found in one geological stratum was typically followed by noticeably different fossils in younger strata. The numerous intermediate forms that one might expect were not seen.

Darwin attributed this phenomenon to gaps in the fossil record. He had no doubt that the intermediate forms had existed. They had simply not been fossilized or had not been found. In Darwin's day, paleontology was in its infancy, and it was only to be expected that there existed numerous important fossils that had simply not yet been found. Nevertheless, Darwin understood that the gappiness of the fossil record was a difficulty; he wrote:

The geological record is extremely imperfect and this fact will to a large extent explain why we do not find interminable varieties, connecting together all the extinct and existing forms of life by the finest graduated steps. He who rejects these views on the nature of the geological record, will rightly reject my whole theory.

But the fossil record has remained full of gaps to the present day. To be sure, some intermediate forms, such as whales and snakes with legs, have been discovered. However, more often than not, new forms are still seen to appear suddenly in geological strata. The slow changes that Darwin's theory of gradualism seems to require are generally not observed. One species may appear to be descended form another, but the fossil record generally does not show how one evolved into another.

Nowadays orthodox Darwinians do not consider the fossil record to be a difficulty. For one thing, fossilization is a rare event. Only a minute fraction of one percent of all living organisms are preserved as fossils. In fact, there exist numerous living species today that have no fossil history. They tend to live in habitats in which fossils are unlikely to be formed. Furthermore, the vast majority of fossils that are created are never discovered. Fossils are frequently destroyed by geological forces and chemical change. Many others are buried so deep in sedimentary rock that they are likely never to be found. Others, which happen to lie in exposed rock strata, erode away long before they are found. Nature simply does not cooperate very well with the wishes of paleontologists.

But this is only part of the story. When we look at the way in which new species are thought to be created, it immediately becomes apparent that a certain amount of gappiness in the fossil record may be inevitable. Remember that speciation, the creation of new species, is thought to be a gradual process. Evolution, as Darwin recognized, does not proceed by sudden leaps. Large mutations do occur in some cases, but they are uncommon and generally fatal. What normally happens is that small mutations occur in part of a population, causing a species to split into two different varieties.

If two slightly different varieties lived in close proximity to one another, speciation would not occur. Instead, the two—very slightly different— varieties would interbreed with one another, and the differences between them would disappear.* If a population of giraffes with necks that are slightly shorter than average bred with one possessing necks that we slightly longer than the norm, a single population of giraffes with necks of normal size would be the likely result.

But suppose that two populations of a single species become geographically separated from one another. Some might migrate over a mountain range during a warm period and find themselves separated from other members of the species when conditions turned colder. Alternatively, two populations might become separated by a desert or live in a lake that splits into several different lakes when conditions become dryer. Then, if orthodox Darwinian theory is correct, they will evolve along separate lines and gradually become distinct from one other. They may adapt themselves to different environmental conditions and thus develop different adaptations. Chance factors may play a role, too. If a small population becomes separated from the remainder of the species, it probably will not be genetically identical. Chance alone will ensure that some genes will be more frequent and others less so.

Now suppose that, at some later time, the two populations come into contact again. If there has been enough time for the two populations to evolve sufficiently, they will now be distinct species and will either not breed or be unable to produce fertile hybrid offspring. If this is the case,

---

*In Chapter 8, we will see that this hypothesis is beginning to be questioned. It has been suggested that it may not always be true. But I will stick to describing the orthodox viewpoint for now.

one of the two species is likely to replace the other, at least in the region where they come into contact. Closely related species generally compete with one another, and the one that is better adapted to the environment is likely to survive while the other becomes extinct.

If something like this happens, then we should expect to see a gap in the fossil record. The gap will be there not because there was any sudden evolutionary change but because one species replaced another. This is especially likely to be the case if the population that became isolated was a small one. Small populations can evolve more rapidly because new, favorable mutations can spread through them more quickly.

If a new species appears "suddenly" in the fossil record, it isn't necessary to conclude that it replaced another species over a period of years, decades, or even centuries. To paleontologists, 100,000 years is a geological instant. The geological record, after all, stretches back over periods of hundreds of millions and billions of years. When rock strata are examined, 100,000 years seems to be nothing more than a moment of geological time. In fact, in most cases it would not be possible to tell whether one stratum was 100,000 years older or 100,000 years younger than another.

What the geological record does show us is that a species living at one time is often replaced by another species living millions of years later. It is perfectly plausible, indeed likely, that the first species that is found did not gradually evolve into the second but was replaced by the latter when it migrated into the region.

Alternatively, the first species may no longer be living in that region when members of the second arrive. When climatic conditions change, species generally do not try to adapt to the new conditions. They migrate. When the climate grows colder, for example, terrestrial plants and animals gradually move south. Plants species can migrate too, because they spread seeds. A plant adapted to warm conditions will naturally propagate seeds in all directions. But if the winters are becoming more severe, those seeds that sprout in more northerly latitudes will be less likely to survive. The opposite will be the case at the southern end of the plant's range.

This phenomenon is known as habitat tracking. Habitat tracking is probably responsible for much of the gappiness we see in fossil beds. Members of one species cease to fossilize there because they now live elsewhere. On the other hand, a daughter species may be adapted to somewhat different conditions and may now find that region quite congenial.

There is ample evidence that habitat tracking has taken place. Lions and hippopotamuses lived 120,000 years ago in what is now England. The climate was warmer then, and they were able to survive quite easily in the climate that existed there at the time. Habitat tracking has also been observed in recent decades as Earth's climate has become slightly warmer. For example, the armadillo and the Virginia opossum have been extending their ranges northward in the United States. Something similar has been happening with various species of birds, insects, and plants.

It appears that the fossil record has exactly the character it should have if Darwin's theories of natural selection and gradualism are correct. There should be gaps for the simple reason that species typically do not remain in the same place over very long periods of time. When one fossil is found above another in layers of sedimentary rock, it is possible, indeed probable, that the second evolved somewhere else. Fossils found at a given geological site will not exhibit a series of small evolutionary gradations if evolution happened elsewhere.

Evolutionary Stasis

Environmental conditions in the oceans do not vary as much as they do on land, and marine organisms often have ranges that are much greater than those of terrestrial animals. Marine organisms are more likely to become fossilized, as well, because there is a greater probability that they will be buried in sediment.

Thus when Niles Eldredge began studying fossils of some long-extinct marine animals called trilobites during the 1960s, he fully expected to find patterns of gradual evolutionary change. The trilobites, which lived from about 540 million to 245 million years ago, dominated the oceans. Numerous fossils had been found, and Eldredge at first didn't think he would experience a great deal of difficulty documenting the slow evolution of one trilobite species into another.

But this is not what he found. He would pick out a trilobite species and find that it changed very little over periods of up to 8 million years. There were some differences between earlier and later specimens, but they didn't seem to be very significant. As far as he could tell, evolution had been going nowhere.

Research projects are supposed to yield results, and Eldredge felt very frustrated at first. He had picked out an area of study, and he wasn't getting anywhere. Then, gradually, he began to realize that he had rediscovered a phenomenon that had been well known to paleontologists who had been contemporary with Darwin. These paleontologists were aware that species were very stable entities. Stasis, not gradual change, was the norm in the fossil record. And, in fact, all five of the paleontologists who had written reviews of Darwin's *On the Origin of Species* upon its appearance in 1859 had pointed out this fact.

So why, then, wasn't the fact of evolutionary stasis generally known? Apparently paleontologists had adopted the orthodox idea of gradual evolutionary change and had held onto it, even when they discovered evidence to the contrary. They had been trying to interpret fossil evidence in terms of accepted evolutionary ideas. They had seen stasis in various evolutionary lineages over and over again and had not realized that they were observing something important.

That they did not do so isn't as surprising as you probably think. Scientists normally interpret the facts they observe in terms of existing theory. If they didn't do this, it would be impossible to do science at all. If a new theory were created to explain every new experimental result or empirical observation, the result would be chaos. So scientists habitually try to fit their results into existing theoretical frameworks. Normally, this works pretty well. Of course there are sometimes anomalous results of one kind or another, but in most cases, they can be safely ignored or explained away. Quantities are sometimes measured inaccurately, and observations can be influenced by extraneous factors.

But sometimes the apparent anomalies aren't that at all. Sometimes they reveal something significant. Sometimes scientists see a phenomenon that accepted theory failed to predict. As Eldredge pondered the evolutionary stasis he had found in his trilobites, he began to wonder if perhaps he hadn't stumbled onto something important. If the stasis he saw was real, it might indicate that accepted ideas about gradual evolution were not entirely correct.

You shouldn't imagine that Eldredge's findings cast any doubt on the idea of natural selection. He had only observed that certain species had remained static for long periods of time. They had certainly evolved from earlier forms, and natural selection was the only thing that could have caused them to do so. On the other hand, the existence of stasis implied that evolution might not always be the gradual process that Darwin had envisioned. And if this was the case, there were other implications. If evolution could, in effect, grind to a halt for long periods of time, then there would have to be other periods during which it proceeded more rapidly.

Gould's Snails

Stephen Jay Gould is well known as a writer on evolutionary biology. He writes a monthly column for the magazine *Natural History* and is the author of numerous best-selling books. But as a scientist, Gould specializes in the study of fossil snails. Now it so happened that, at about the same time Eldredge was seeing stasis in species of trilobites, Gould was observing the same thing in the fossil snails he studied. Like Eldredge, he found that new species made abrupt appearances in the fossil record and then remained more or less the same for long periods of geological time.

In 1971, Gould was asked to contribute a paper to a book that was to be titled *Models in Paleobiology*. He was asked to write about speciation. This was not the topic that he would have preferred, so he asked Eldredge to write the paper with him. As paleontologists, Gould and Eldredge were familiar with one another's work, and they realized that they had both observed something they could write about in the paper.

The two paleontologists wrote a paper called "Punctuated Equilibria: An Alternative to Phyletic Gradualism." It was published in 1972. In it, Gould and Eldredge argued that new species did not arise "from the slow and steady transformation of entire populations." On the contrary, most evolutionary change took place when new species arose "very rapidly in small, peripherally isolated populations." They went on to claim that "the great expectation of insensibly graded fossil sequences was a chimera." The gaps in the fossil record, they said, recorded something that was very real.

According to Eldredge's and Gould's theory, which is known as the theory of "punctuated equilibrium," most evolutionary change takes place when new species are created. Small, isolated populations adapt themselves to local conditions, branching off from the parent species as they do. For a short time, natural selection causes them to evolve rapidly.

And then nothing much more happens. According to Eldredge and Gould, evolution was not something that took place over periods of many millions of years. Adaptation over periods of tens of thousands of years was more likely. When the fossil record was examined, the appearance of sudden evolutionary changes was evident. As I pointed out previously, a period of 100,000 years is a "geological instant" compared to the many of millions of years represented by the rock strata that scientists study. Thus a period of tens of thousands of years would be only a fraction of an instant.

Eldredge's and Gould's findings seemed to imply that, most of the time, natural selection acted to keep species stable. It produced evolutionary change only when a new species arose and had to adapt to local conditions. After all, if a species was already adapted to its environment, further change was likely only to lower the average fitness of the individuals that made up the species. If Eldredge and Gould are correct, then natural selection, too, seems to follow the principle "If it ain't broke, don't fix it."

According to Gould, it was Eldredge who developed most of the ideas associated with the theory of punctuated equilibrium. If this is true, and Gould is not just being modest, he has nevertheless contributed significantly to the development of the theory. Both Gould and Eldredge have thought and written about the implications of the theory, and Gould has been its most vocal defender.

However, it is Eldredge who has discussed the theory at greater length. In his book *Reinventing Darwin* (John Wiley, 1995), he discussed his and Gould's ideas in great detail and explained what some of the implications of punctuated equilibrium might be. According to Eldredge, the fact that evolution has proceeded in short bursts, rather than as a slow and steady process, implies that it is somewhat more complex than the orthodox Darwinists claim. There is more to it than the action of natural selection on genes or on individuals.

Eldredge and Gould believe that entire species play a role in evolution. As Eldredge puts it, "The species that happen to develop adaptive change in one particular direction are the ones that tend to survive, to produce further species, thus handing down to descendants those very features that conveyed success on themselves." In other words, species are packages of genetic information subject to something analogous to the action of natural selection on individuals. According to Eldredge and Gould, natural selection acting on individual organisms may be the most important factor in evolution, but it is not the only one. The differential survival of species, which Gould and Eldredge call species sorting, also has to be taken into account.

If there were no such thing as mutation, evolution could not happen. Random mutations occur in the genes of all organisms, and they create the variation on which natural selection operates. Natural selection eliminates the harmful mutations while preserving the beneficial ones. The beneficial mutations are thus passed on to successive generations. As they accumulate in a population, adaptive changes take place. According to Gould and Eldredge, there is nothing corresponding to mutation at the level of a species. Speciation is not analogous to mutation, and species sorting is not a kind of natural selection at a higher level. Natural selection and species sorting, they say, work together to produce evolutionary change.

Eldredge and Gould speculate that new species commonly arise when a population of the species that lives in a marginal habitat becomes isolated from the remainder of the species. If this happens, natural selection will cause the population to become adapted to its particular environment. When it does, the habitat will no longer be suboptimal. The population will evolve into a new species that is adapted to different conditions. Thus speciation itself can be seen as a source of change. Natural selection is not the only factor involved. If the population had not become isolated in an environment to which the species was not well adapted, no evolutionary change would have taken place.

The number of species cannot multiply without limit. As I have noted previously, very similar species will compete for resources when they live in the same environment. This will lead to a sorting-out process that will ensure that only some species survive.

Eldredge's and Gould's theory can therefore be broken down into four parts:

1. Most of the time, natural selection does not cause significant evolutionary change. On the contrary, natural selection acts to keep species stable by weeding out individuals that deviate too greatly from the norm. If a species does not quickly become extinct, it may remain essentially unchanged for many millions of years.

2. Evolutionary change takes place in short bursts. These bursts have a duration that can be measured in tens of thousands rather than millions of years.

3. Evolutionary change takes place because populations of a species sometimes become isolated in habitats that are initially less than ideal. Since natural selection acts to improve the fit between organisms and their environments, the population will evolve into a new species.

4. Not all new species can survive. There will be a kind of sorting out, called species sorting, that will cause some of them to become extinct, while other perpetuate themselves and give birth to new species. Species are packages of genetic information. Some of them perpetuate themselves, while others do not. Species sorting, therefore, is a factor in long-term trends in evolution.

The idea that natural selection was not the sole source of evolutionary change did not sit well with the orthodox Darwinians. For example, the British geneticist J. R. G. Turner derided the theory of punctuated equilibrium as a theory of "evolution by jerks," while the British paleontologist Lionel Beverly Halstead characterized it as a Marxist tract masquerading as a serious scientific idea. The idea seems to have been that describing evolutionary history as something that had been punctuated with sudden "revolutions" rather than slow "progress" betrayed a Marxist outlook.

Turner's and Halstead's criticisms can be easily dismissed. Turner really contributed nothing to the debate over punctuated equilibrium beyond coining a memorable phrase. And Halstead's charges of Marxism can hardly be taken seriously. Neither Eldredge nor Gould is a Marxist, although Gould has made it clear that his political sympathies lie with the left. On the other hand, the criticisms of John Maynard Smith have been more substantial. For example, in 1987, Maynard Smith wrote a commentary in the British journal *Nature* in which he discussed a paleontological study of trilobites by paleontologist Peter Sheldon that was published in the same issue. In his commentary, Maynard Smith did not deny the existence of stasis, but he did maintain that the study provided evidence that gradual evolutionary change did appear. In the course of his article, which was titled, "Darwinism Stays Unpunctured," he criticized the idea of species sorting. In Maynard Smith's mind, it was still obvious that natural selection was the source of all evolutionary change.

Naturally Eldredge and Gould took issue with what Maynard Smith had said. However, the debate came to no definite conclusions. Eldredge and Gould saw stasis in Sheldon's study, where Maynard Smith saw gradual change. Maynard Smith accused Gould and Eldredge of promulgating species sorting as an alternative to natural selection. The two Americans replied that they had not. Species sorting determined the eventual fate of adaptations that had evolved. Natural selection was, of course, responsible for their appearance in the first place.

Neither side won the debate. The orthodox Darwinians continued to maintain that, contrary to what Eldredge and Gould said, all the observed patterns in evolutionary history were the result of natural selection within local populations. Both individuals and species, they said, were packages of genes. And there existed nothing besides natural selection that could determine their genetic makeup. Eldredge and Gould, on the other hand, continued to maintain that the orthodox Darwinists' reductionist methods failed to explain everything. The orthodox view embodied a kind of blindness. Or, as Eldredge put it, "ultra-Darwinians know the principles of what moves the car, but they cannot imagine the road conditions under which it is driven."

"Puncturing Punctuationism"

Richard Dawkins devoted an entire chapter of his 1986 book *The Blind Watchmaker* to criticisms of the theory of punctuated equilibrium. Dawkins began this chapter by explaining what punctuated equilibrium was not. Eldredge and Gould, he said, were not suggesting that evolutionary changes could take place within a single generation. They were no more saltationists than the orthodox Darwinians were. In fact, they were just as gradualist as the biologists they were criticizing. Evolutionary change that took place in tens of thousands of years was still slow by human standards, Dawkins said. Finally, Eldredge and Gould were not attacking the idea that evolution always proceeded at a constant rate. No one seriously believed that proposition.

In other words, Dawkins said, there was nothing very revolutionary about the Eldredge–Gould theory at all. "The theory of punctuated equilibrium lies firmly within the neo-Darwinian synthesis," he claimed. (When Dawkins uses the words "neo-Darwinian synthesis," he is referring to what I have been calling "orthodox Darwinism.") In other words, one could easily accept the idea that evolution sometimes proceeded in bursts because these bursts were just another kind of gradual change brought on by natural selection.

Punctuated equilibrium was a relatively unimportant addition to orthodox theory, Dawkins went on. Unfortunately, it had garnered a lot of publicity, giving the impression that Darwinism was being challenged. This was unfortunate. There were far too many people in the world, Dawkins said, "who desperately want not to have to believe in Darwinism." As a result, if any reputable scholar even hinted that he was criticizing some detail of current Darwinian theory, his objections would be reported in the media and blown up out of all proportion. Dawkins hinted that Eldredge and Gould, by making their ideas seem more revolutionary than they really were, had inadvertently given the message that something was wrong with Darwinism. He then quoted the editor of *Biblical Creation* as saying:

> It is undeniable that the credibility of our religious and scientific position has been greatly strengthened by the recent lapse in neo-Darwinian morale. And this is something we must exploit to the full.

Dawkins admitted that Eldredge and Gould had "shouted complaints" at the misuse of their ideas. However, Dawkins noted, this part of their message hadn't gotten across. It simply wasn't news. It would take time, Dawkins said, to undo the damage done by the "overblown rhetoric" that accompanied the reporting of Eldredge's and Gould's ideas. But, he went on, he was sure that it would be undone and that punctuated equilibrium would come to be seen as what it really was, a "minor wrinkle on the surface of neo-Darwinian theory."

Dawkins went on to criticize the idea of species sorting in the following chapter. But he did not discuss the idea at great length He noted simply that he was skeptical about the idea. In most cases, complex adaptations were not properties of species, he said, they were properties of individuals. And if a species went extinct, that was presumably because all the individuals in the species were poorly adapted. In his view, species did not have any significant traits that were not possessed by the individuals that made up the species. "It is hard to think of reasons why survivability should be decoupled from the sum of the survivabilities of the individual members of the species."

In other words, according to Dawkins, Eldredge and Gould had not hit on a phenomenon of any importance. Individuals lived or died, individuals either passed their genes on to the next generation or they did not. It was this that caused evolution to take place. The fact that individuals were grouped together in species, according to Dawkins, had little or no effect on evolutionary patterns. Some species might be more likely to avoid extinction than others, but this wasn't due to any special species character. What a theory of evolution should do, Dawkins maintained, was to explain "well-designed mechanisms like hearts, hands, eyes, and echolocation [the sonar system of bats]." Species selection, or species sorting, couldn't do this.

"A Hopeful Monster"

Daniel Dennett repeated many of Dawkins's criticisms in *Darwin's Dangerous Idea,* which was published some nine years later. He began his discussion in a section of his book titled "Punctuated Equilibrium: A Hopeful Monster" by emphasizing, as Dawkins did, that punctuated equilibrium was as much a gradualist theory as orthodox Darwinism was. But his major criticism seemed to be that Gould had claimed "revolutionary" status for a theory that wasn't revolutionary at all. When Gould was criticized for this, Dennett said, Gould had "backpedaled hard, offering repeated denials that he had ever done anything so outrageous."

Gould, Dennett said, had changed his mind several times about what it was that he and Eldredge were claiming. Gould had flirted with the idea of saltational change at one point, and had proclaimed the death of gradualism at another. But, in fact, Dennett said, Gould was a gradualist himself.

The idea of evolutionary stasis, Dennett went on, was really not much of a contribution to evolutionary theory and was not even a common pattern in evolutionary history. Most evolutionary lineages, Dennett said, died out before they had a chance to exhibit stasis. It was only possible to "see" a species when it represented something stable in the fossil record. If it hadn't had some degree of stasis, it simply not have been noticed. Dennett compared Eldredge's and Gould's discovery of stasis to the observation that all droughts last longer than a week. We never notice that a drought exists unless it persists for long enough a time to be remarked on.

In their 1972 paper, Eldredge and Gould had suggested that the cause of stasis might be an inborn genetic conservatism, that species might actively resist adaptive change. Dennett devoted several pages to a critique of this idea, pointing out that, if this were the case, it should be difficult to breed desired traits in domestic animals. But the breeding of domestic animals was relatively easy, Dennett pointed out, at least at first. Breeding caused their characteristics to change rapidly over a small number of generations; it only became difficult when a breed reached the point where most natural variations had been bred out of it.

Dennett concluded that, when one looked at the theory of punctuated equilibrium closely, it turned out that there was little to the idea. Eldredge and Gould had made a "conservative correction to an illusion to which orthodox Darwinists had succumbed." Paleontologists had thought that natural selection should create a fossil record containing a lot of intermediate forms. Eldredge and Gould had corrected this mistake. Beyond that, they had contributed little or nothing.

Dennett then went on to the idea of species sorting. He didn't reject it out of hand, as Dawkins had, saying instead, "The relative importance of species selection of the sort Gould is now proposing has not yet been determined." Species sorting might play some role in evolution, he said. It might be a previously unrecognized mechanism "built out of standard, orthodox mechanisms." In other words, it was not yet possible to decide whether the idea of species sorting was correct. But if it turned out that it was, then it was something that depended on natural selection acting in the usual manner.

However, Dennett went on, he had suspected all along that Gould had been trying to introduce non-Darwinian ideas into evolutionary theory. Gould, he claimed, had been following a program of attempting to cast doubt on the universal validity of Darwin's idea of natural selection. What Gould really objected to was the image of a *"predictable, mindless trudge"* (Dennett's italics) toward adaptation. Gould's real target, Dennett charged again, was Darwin's dangerous idea itself.

Defending Punctuationism

Eldredge has shied away from engaging in public controversy over punctuated equilibrium. However, he has continued to write about the theory. In his book *The Pattern of Evolution* (W. H. Freeman and Company, 1998), he responded to some of Dennett's criticisms without mentioning the latter by name. Eldredge admitted that his and Gould's initial idea of inborn genetic resistance to change never gained much popularity. But this was only one of several ideas that he and Gould had explored, he said, adding that he now thought that a more likely explanation of stasis is that "stabilizing natural selection will be the norm even as environmental conditions change." As long as a species is able to adapt to changing conditions by moving and tracking familiar habitats, it is able to remain more or less the same as the environment keeps changing. This is an idea we have met before: habitat tracking may explain stasis.

Stephen Jay Gould has been more vocal. In "The Pleasures of Pluralism," the second of his two articles in *The New York Review of Books,* Gould turned to Dennett's criticisms of punctuated equilibrium after replying to some of Dennett's other charges. According to Gould, it was Dennett's position that Gould had begun his revolution-mongering in the 1972 paper on punctuated equilibrium. In reality, Gould said, he and Eldredge had denied any revolutionary intent. In response to Dennett's charge that "Gould has several times changed his mind about what he and Eldredge were claiming," Gould replied that they had simply "altered and expanded the theory in recognizing further implications and dropping untenable conclusions—as any active and interesting formulation of a scientific theory must continually do." But, said Gould, Dennett had cared little about this method of scientific work. He had been interested only in charging Gould with "flimflam and backpedaling."

Gould then replied to Dennett's charge that he and Eldredge had flirted with the idea of saltational, non-Darwinian change. "Of course Dennett cannot quote us on this," Gould said, "because we never said such a thing." The theory, said Gould, had indeed been mischaracterized in this manner, and he and Eldredge had issued rebuttals. Nevertheless, Dennett seemed to want to blame Gould for the confusion about what he and Eldredge had really said.

Gould accused Dennett of willfully misreading and misrepresenting the things he had said. Dennett repaid the complement in a letter to the editor of *The New York Review,* published on August 14, 1997. Dennett insisted that he had been correct when he said that Gould "had tried on saltationism and then abandoned it." On the contrary, Dennett claimed, this interpretation was "standard fare, widely accepted in the field." Dennett repeated his claim that punctuated equilibrium had invoked "distinctly non-Darwinian mechanisms for stasis and change."

Orthodox Darwinism, which Gould "propagandistically elides into 'Darwinian fundamentalism'," Dennett insisted, was alive and well, despite Gould's claims to the contrary. When Gould suggested that it was not, he was misleading the public. Dennett went on:

> Let me say a word about "Darwinian fundamentalism." Nonsense. I do not espouse the preposterous views Gould attributes to this mythic creed. Gould labors to create a caricature of the "strict" adaptationist, a type that occurs nowhere in nature and is disavowed at length by me. In fact, the passage which Gould uses to anchor this fantasy is a misquotation. . . . What is amazing is that Gould wrests this quotation from the very section in which I attempt to undo the travesty of Gould's previous attempts over the years to caricature adaptationist thinking.

"I am sorry it has come to this," Dennett concluded. "I have tried hard to get ⌊Gould⌋ to stop misrepresenting the works that he disapproved of, to clarify his position, and to disavow the misconstruals that are so often expressed by nonbiologists citing him as their authority." But Gould had not deigned to do this, Dennett said. Instead he had chosen to "turn up the volume of his vituperation."

At this point, Gould refused to continue the argument, saying that Dennett had written a "singularly contentless commentary which reminded me of this motto and its corollary, 'When you have nothing to say, say it louder.'" Dennett had offered "absolutely no intellectual response," Gould said. "He only avers that he can support a false claim he had made about punctuated equilibrium with a bevy of quotations that he didn't bother to use." In the end, Gould went on, Dennett's response had been nothing but "bluster."

But Gould found himself unable to stop there. He found it necessary to complain again about Dennett's attempts to link his intellectual positions with his supposed political and religious views. Then Gould once again pointed out that Dennett had taken his quotation of an African-American folk song to be a Biblical citation. To some readers, it must have seemed an odd note on which to conclude the argument.

An Odd Scientific Controversy

Scientific controversies generally revolve around questions of the correct interpretation of data or around the validity of new theoretical ideas. The controversy between Gould and his colleagues, on one hand, and Dawkins, Dennett, and Maynard Smith on the other, is nothing like that at all. The orthodox Darwinians admit that most of the ideas advanced by Gould and company are probably correct. Rather than argue against them, they say that these ideas are really of only minor importance. Gould and his colleagues, they say, have added only a few "minor wrinkles" to orthodox theory.

Gould, on the other hand, maintains that these ideas are important and calls his opponents "Darwinian fundamentalists" because they choose to emphasize the importance of natural selection above all else. Meanwhile, members of both camps accuse their opponents of misunderstanding and misrepresenting their views.

It is difficult to see how a controversy of this nature can be resolved. The two camps are arguing, for the most part, not about interpretations of scientific fact but rather about their different ways of seeing patterns in evolution. The orthodox Darwinians believe that natural selection is by far the most important factor and that others are of minor importance. Gould and his colleagues prefer to look at evolutionary patterns in a more "pluralistic" manner. They do not deny the importance of natural selection, but they maintain that other important factors are at work too.

Scientific controversies are not always resolved. Sometimes they are simply forgotten. New generations of scientists sometimes pay little attention to the arguments of their predecessors. They may have their own ways of thinking, or they may come to feel that the outlook of one camp or the other seems more natural.

If controversies are resolved, it is usually because important new evidence is discovered that favors one side or another. To see whether that is likely to happen in this case, it might be well to restate the ideas that Gould and his colleagues have expressed and to speculate about the kinds of evidence that might support these ideas.

I think the ideas advocated by Gould and company can be broken down into three categories: (1) contingency, (2) spandrels, and (3) punctuated equilibrium.

Contingency

Gould is fond of pointing out that contingency plays an important role in evolution. If we could roll the clock back 3.5 billion years and allow evolution to start again from the beginning, it is likely that different kinds of organisms would evolve. As Gould points out, chance has played an important role in the evolution of life on Earth. For example, there have been numerous mass extinctions, and chance has played a role in determining which species would and would not survive. The dinosaurs did not become extinct because they were less well adapted than the mammals. They died out because they were unlucky. Natural selection could not foresee that Earth would collide with an asteroid. It adapted the dinosaurs to existing conditions, and it did this superbly. The mammals may have survived *because* they were unable to compete against the dinosaurs in the ecological niches the latter occupied. For example, they might have continued to flourish because they were small. Perhaps their small size gave them some kind of advantage during the years following the collision.

Approximately 96 percent of all living species perished in the great Permian extinction, which preceded the extinction in which the dinosaurs perished. Most marine species were wiped out. But the clams survived. Now there seems to be nothing magical about being a clam. Apparently there was something about their makeup that caused them to be better adapted to the changed conditions than other species. But this was most likely to have been simply the result of chance. If the Permian extinction had had a different character (scientists are not quite sure what caused it, but obviously environmental conditions changed in such a way as to make survival difficult) or had happened millions of years earlier or later, the clams might have perished while other species went on.

If the patterns of the mass extinctions that Earth has experienced had been a little different, Earth might now be populated by organisms entirely unlike those living today. Perhaps intelligent life would never have evolved. Alternatively, there might be two or more intelligent species, or intelligent species might have some form that would seem exceedingly bizarre to us. For example, it is not beyond the realm of possibility that Earth might have come to be dominated by intelligent reptiles or even intelligent giant worms, for that matter.

Extinctions are the most dramatic of the chance events that have altered the course of evolution. Sometimes there is more than one way in which an organism can become better adapted, and natural selection may favor one path over the other simply by chance. Certain combinations of genes may happen to come together, and this may be enough to cause evolution to follow one path rather than another. For example, like all tetrapods, we have limbs with five digits. There is no obvious reason why four or seven would not have worked as well. But once the number five was chosen, that particular trait was locked in. Most probably it became locked in simply because things happened that way, not because there was anything special about the number 5. If we're going to have digits at all, there has to be some number of them. It is reasonable to conclude that the number of digits is five simply because, at one time, one of our evolutionary ancestors wound up with five and passed on that trait to all its descendents.

There are probably other ways the body plans of tetrapods or of vertebrates in general might have turned out to be different from what they are. Some human characteristics, for example, appear to be pretty arbitrary. There seems to be no particular reason why we should have to have the precise number of ribs or the number of neck vertebrae that we do. Presumably other configurations would have worked as well. We are as we are because our evolutionary ancestors just happened to have body plans of a particular type.

So there apparently should be a great deal of contingency in evolutionary patterns. No one denies this. The orthodox Darwinists would, in fact, agree with most of what Gould says about this matter. They simply don't think it is an issue of any great significance. *Of course* evolution could have happened differently, they say, but so what? In their view, chance factors may indeed play a role, but it is still natural selection that does all the work.

I personally don't see how the discovery of any new evidence could settle this argument. In this case, Gould and the orthodox Darwinians are not arguing about facts. They simply have different points of view. For that matter, their respective points of view are not really so incompatible. Gould may find it necessary to explain the idea of contingency when he writes for a popular audience. But I suspect that, for most biological scientists, the idea is really nothing new. I'm a little surprised that Gould places as much emphasis on the idea as he does.

Spandrels

Similarly, the orthodox Darwinists generally don't claim that Gould's and Lewontin's theory of spandrels is wrong. They simply deny that the theory is very important. They don't deny that certain traits of an organism may have been created as a by-product of traits engineered by natural selection. But they see natural selection as something that is always at work and prefer to think that it is natural selection that adapts spandrels to any adaptive purpose for which they might be used.

The scientists who work in the relatively new discipline of evolutionary psychology see the human brain as something that can be broken down into distinct "modules" that evolved for adaptive reasons. Gould, on the other hand, says that the brain is full of spandrels that were not created by natural selection for any particular adaptive purpose. Since the question of the significance of spandrels takes on a more concrete form in the controversy between Gould and the evolutionary psychologists, it might be best to defer a fuller discussion until later.

However, it is possible to say that it is hard to see how any new empirical evidence can settle the question of the importance of spandrels in the evolution of physical traits. As Gould and Lewontin pointed out in the article in which the concept was introduced, we often cannot tell why a certain trait originally evolved. In many cases, we can only invent plausible stories that may or may not be true. No one doubts that the swiftness of antelopes evolved to give them a better chance of escaping from predators. But many other traits could be explained in a number of different ways. And it is often impossible to tell whether these traits are spandrels or not. Sometimes the question of why a given trait exists is among the most difficult ones in biology.

Punctuated Equilibrium

Eldredge's and Gould's theory of punctuated equilibrium is considered to be their most important contribution to evolutionary theory. Again, the orthodox Darwinists tend to downplay its importance, saying that evolution in short "bursts" is still gradual, and that Gould and Eldredge did little more than correct an impression, which had been common among paleontologists, that gradual change should always be occurring.

But there is more to punctuated equilibrium than that. After Eldredge and Gould concluded that most evolutionary change took place when new species were created, they speculated further about why this should be so and about the roles entire species might play in evolution. This led them to the idea of species sorting. According to Gould and Eldredge, species are packages of genetic information and the survival or extinction of particular genetic packages can be important. This is disputed by the orthodox Darwinists who doubt that there are any evolutionary processes of importance above the level of natural selection.

There seems to be a real argument here. It seems at first that it might be a difficult one to settle. After all, the evolution of species is something that happens on time scales that are long compared to the human life span. Scientists can observe natural selection at work. But, with a few rare exceptions, the evolution of new species cannot be observed. And if species sorting does play a role, it would be difficult or impossible to observe, too.

An important ingredient of species sorting is extinction, and extinction does not take place overnight, either—except in cases where human activities cause a species to very suddenly become extinct. Species usually do not die out in a day or even over a period of centuries. It is thought that the process of extinction may sometimes take millions of years. A species or a group of species may gradually decline in numbers over a long period of time. It is not even certain that the dinosaurs perished quickly. The collision with an asteroid 70 millions years ago may only have hastened a decline that had already been going on for millions of years.

Nevertheless, it is conceivable that new evidence that has some bearing on punctuated equilibrium or species sorting may turn up. In fact, some may already have been discovered. There is new evidence that natural selection may sometimes proceed at a faster rate than had previously been thought possible. If this evidence is confirmed, it could very well turn out to have a bearing on Eldredge's and Gould's ideas.

But before I look at this evidence, I would like to discuss some other controversies that are going on in the field of evolutionary biology. Stephen Jay Gould is not the only evolutionist who has presented new ideas, and his arguments with the orthodox Darwinians are not the only ones that are being heard. For example, some of the scientists who work in the sciences of complexity believe that they have made discoveries that are relevant to an understanding of the evolutionary process.

The Sciences of Complexity

Ants have simple nervous systems. The brain of an ant is not very complicated, and individual ants should probably be regarded as unconscious automatons. Interactions between ants are not very complex either. They signal to one another in only about a half dozen different ways. And yet the behavior of colonies of ants, which may contain anywhere between 5,000 and 20 million individuals, can be astounding. Some species of ants make slaves of members of other ant species. Other species have learned to be "farmers." They cultivate the fungi that they eat. Yet others "milk" aphids for the nutrients they need to survive. Ants are able to construct elaborate nests, defend their nests, and engage in efficient foraging behavior. In general, ants do numerous different things that the hypothetical visitor from another planet, who knew only the details of their anatomy, would never expect.

Other social insects that exhibit complex group behavior include termites and bees. Some species of termite, for example, build wedge-shaped nests that are always oriented in a north–south direction. Others build nests that have covered runways leading to them. Some species have even built arch-shaped structures. And yet individual termites are also animals with limited brain capacity. They cannot possibly be aware of compass directions or be capable of knowing that runways would be advantageous. The behavior of a termite colony can no more be predicted from a knowledge of the behavior and physiology of an individual termite than the behavior of an ant colony can be predicted by someone who knows only what individual ants are like.

The behavior of large groups of organisms often exhibits properties that are not seen in individuals. A similar phenomenon can be observed in certain kinds of inanimate objects. "Collective" behavior can be seen in something as simple as a bathtub that has been filled with water. When the plug is pulled, a vortex spontaneously forms above the drain as the water swirls down it. You may have heard that the water will swirl one way in the northern hemisphere and the other way in the southern. If you have, don't believe it. The water may swirl down the drain in either direction. The motion of the water that is induced when the plug is pulled will set it going one way or the other. A bathtub full of water is something like a pencil that has been balanced on its end. Small perturbations—air movements will do this—will cause it to fall in one direction or another, and it is not possible to predict beforehand in which direction it will fall.

The creation of a vortex in a bathtub cannot be predicted from a knowledge of the properties of the molecules that make up the water. This does not mean that scientists do not understand this phenomenon. Vortex motion can be described by the laws of fluid dynamics. However, the laws of fluid dynamics cannot be derived from the laws that describe the behavior of the water's basic components. In this case, a reductionist approach simply does not work.

The behavior of social insects and bathtubs full of water are two examples of a principle that has motivated scientists from numerous different fields to begin looking at certain kinds of problems in a new and different way. This principle can be stated very simply: "Complex systems have emergent properties." In other words, a system of sufficient complexity, whether it is a collection of organisms, of inanimate molecules, or of anything else, will typically have properties that cannot be explained by breaking the system down into its elements. Another way of saying the same thing would be to say, "Complex systems are self-organizing." That is, when a system becomes sufficiently complex, order (e.g., a vortex, complicated social behavior) will spontaneously appear.

Emergent properties can be seen almost everywhere, in avalanches, in air turbulence, in weather patterns, in the formation of spiral galaxies, in living cells, in ecological systems, and in human societies. Human cultures have numerous characteristics that could never be predicted from a knowledge of the psychology of an individual. When we interact with one another, new and complex kinds of behavior appear, just as it does when ants interact.

As recently as several decades ago there seemed to be little hope of understanding the appearance of emergent properties in complex systems. Their presence could be noted. But if they could not be analyzed in a reductionist manner, they could not really be understood. But then something happened that changed all that: the development of modern high-speed computers. When such computers became available, it became possible to create models of complex systems. The behavior of these models could be studied in computer simulations, and insights could be gained into processes that took place in the real world. For example, there now exist models of galaxy formation that have allowed scientists to better understand how certain characteristics of galaxies come about. Similarly, there are models for studying the behavior of social insects. Simulations of hives of bees have been set up inside computers. Computer simulations of ecological systems, the behavior of stock markets, traffic congestion, and many other complex phenomena have also been created.

Most, but not all, of the work done in the sciences of complexity involves computer simulations. In some cases, complex systems can also be studied in the laboratory. For example, two different groups of scientists at the Scripps Research Institute in La Jolla, California, have created collections of molecules that exhibit some of the characteristics of living organisms. They want to see what emergent properties will appear as these collections increase in complexity. If their work is successful, they are likely to gain some insights into the origin of life.

The methods of the sciences of complexity can be applied to research in numerous different fields of science. You probably won't be surprised, therefore, if I tell you that some of it relates to questions concerning the evolution of life. Living organisms, after all, are complex systems, too. They have complex genetic codes, and the components of living cells interact with one another in complex ways. Evolving species too can be viewed as complex systems. This suggests that it might be possible to study certain aspects of evolution by creating computer models of the interactions between genes or of evolving species.

Genetic Networks

Very few traits are caused by the action of a single gene. Human height, for example, is the product of a number of different genes. Most other human characteristics are also the result of a number of genes acting in concert. And there are genes that do not directly code for traits at all. Genes interact with one another. For example, they can turn one another on and off. Some produce proteins that inhibit the production of proteins by other genes. If this kind of inhibition did not exist, there could not be so many different types of human cell. Every cell has the same genetic makeup. Since their differences are caused by the actions of different genes, certain genes must be prevented from expressing themselves in each type. For example, a skin cell is different from a liver cell not because its genetic makeup is different but because different sets of genes are prevented from contributing to the cell's makeup.

So it is generally not possible to find a gene that is "for" a certain trait. Most traits are produced by networks of genes, and a single gene may be part of more than one network. This sometimes causes traits to be linked together in seemingly odd ways. For example, as Darwin himself noted, white cats are often deaf.

There are approximately 100,000 genes in the human genome. The simplest bacteria have many hundreds. Since the number of possible interactions between genes is much greater than these numbers, the genome of any organism must be regarded as a complex system. Consequently, it would be hopeless to try to use reductionist methods to understand all the complex workings of the human genome. It is often not possible to tell what causes a particular gene to be switched on or off or precisely what the relationships between large numbers of genes are. It is sometimes difficult to determine what as few as five functioning objects that are interconnected with one another will do. Thus it may never be possible to determine everything that 100,000 interconnected genes will give rise to.

Hence it seems reasonable to think that work in the sciences of complexity might shed some light on some of the problems of molecular biology. This is precisely what Stuart Kauffman, a theoretical biologist at the Santa Fe Institute in New Mexico, has attempted to do. Kauffman has constructed computer models relating to the origin of life, to interconnections between genes, and to the evolvability of genomes that are interconnected in various distinct ways.

When Kauffman began thinking about genetic networks, he realized that microbiologists could, at least in theory, look at a single gene and find out what other genes interacted with it. However, he didn't think that this kind of reductionist approach would ever be able to fill in all the details of what was going on. He therefore decided to think of the human genome as a complex system with 100,000 components.

Now Kauffman didn't say to himself, "Well, a genome is a complex system. Let's look at the system as a whole and try to see what its emergent properties are." Kauffman's first contact with the problems of genetics was in the 1960s, when he was a medical student. At the time, the field we call the sciences of complexity did not yet exist, and ideas about emergent properties had not yet been well articulated. However, Kauffman did think that there were ways of dealing with the problems he had begun to think about. He concluded that, before one could hope to answer questions about complex gene networks, it was first necessary to understand the characteristic behavior of large networks made up of any kind of component. In other words, he suspected that complex networks would possess certain properties that did not depend on what they were made of. These properties should be the same whether the components were genes or buttons tied together with threads or abstract entities a computer model.

Today, Kauffman's idea does not seem to be especially striking. Scientists know very well that complex systems made up of different kinds of components often give rise to the same types of emergent properties. For example, an avalanche on a snow-covered mountain exhibits behavior like that of an avalanche in a pile of sand, and both can be modeled in computers. Similarly, the emergent properties exhibited by a flock of birds are very much like those seen in a herd of stampeding animals, and they can be modeled in similar ways. However, when Kauffman began thinking about interactions between genes, his idea that a genetic system might have properties that do not depend on the nature of its components would surely have been considered heretical if it had been widely known. At the time, molecular biologists were committed to a reductionist program (in fact the majority of them still are), and they would have laughed at the idea that genetic networks might have properties that did not depend on the precise nature of genes.

Kauffman believed that a genetic network would be self-organizing, that the network would have properties that had nothing to do with natural selection. Certain types of order should spontaneously arise, he thought, and it should be possible to determine what they are. Once this order appeared, evolution would make use of it. But this order would not be initially created by natural selection.

At the time that Kauffman began pursuing these ideas, he was a medical student and naturally had no grant funding. Thus he had to borrow money from his medical school in order to buy the computer time he needed to create models that would test his ideas. At the time—the mid-1960s—mainframe computers were slower and much less powerful than desktop computers are today, and data were fed into them on punch cards. Thus Kauffman had to work with very simple models of gene networks if he was to have any hope at all obtaining results. If his models were too complex, then it might take years for the relatively primitive computers of the day to work out all the possibilities. And of course there was always a chance that Kauffman would discover nothing at all. In that case, the money he spent would be wasted. However, he thought the idea was a good one, and he went ahead.

When Kauffman ran his computer simulations, he found that spontaneous order did seem to appear in his gene network models. But the work he did as a medical student was only a beginning. It was only in later years, when he had an academic position and access to funding (including a "genius" grant from the MacArthur Foundation) and faster computers that he was able to work out his ideas fully. When he did this, he obtained results that exceeded his initial expectations.

Genomes as Complex Systems

Like Gould and Eldredge, Kauffman believes that, although natural selection is important, it is not the sole cause of evolutionary change. "The natural history of life," he says, "is some form of marriage between self-organization and selection." But unlike Gould and Eldredge, Kauffman doesn't speculate about such higher-level phenomena as species sorting. He looks in the opposite direction and tries to see self-organization at the genetic level.

As we have seen, one gene can influence the action of others in a number of different ways. Kauffman does not try to take this into account. In his model, genes can only turn one another on or off. They have no other interactions. Making this kind of simplifying assumption is quite a legitimate procedure. Scientists often approach problems in this manner. If they are successful in creating theories that describe a simplified case, then the theory can be extended to more complicated circumstances later. For example, when Isaac Newton was trying to see if his law of gravitation would describe planetary motion, he initially considered the case of a single planet orbiting the sun. He did not try to take the gravitational influences of all the other planets into account. He simply assumed that these effects would be small compared to the effects of the sun's gravity. As a result, his theory was extraordinarily successful. And Newton's successors worked out the details of situations where the gravitation pull exerted by a number of different bodies had to be taken into account.

Kauffman began by thinking about what the consequences might be if the number of inputs to each gene varied. There is no reason to think that each gene must be influenced by only one other, after all. The protein synthesizing functions of gene A might be turned on or off not only by gene B but also by genes C, D, E, and so on. Kauffman saw that numerous different kinds of gene networks were possible. For example, it would be theoretically possible to have a network in which no gene had any influence on the functioning of any other. And it would be possible to have a network in which every gene influenced every other. In the first case, there would be no interactions. In the second case, there would be a very dense network of effects.

Kauffman didn't consider the case where there was no interaction between genes. It was obvious that nothing of interest would be seen there. He began by looking at a situation where each gene had one input. He found that in such a case, the system would exhibit a tendency to "freeze up." Genes would tend to fall into states where they were permanently on or permanently off. The situation was not very different from one in which there was no interaction between genes at all. It could not be a model of what went on inside a real cell.

On the other hand, when the number of inputs per gene was three, four, five, or greater, the behavior of the system became chaotic. The system would pass through an enormous different number of states without ever repeating itself. Here, "state" means any configuration in which certain genes are on and others are off. For example, consider a very small system (much smaller than those that Kauffman considered) of three genes labeled A, B, and C. A on, B off, C off would be one state. A off, B on, C off would be another, and A on, B on, C off would be a third. There are six possible states in all.

Kauffman concluded that this kind of chaotic activity could not be a good representation of gene activity within a cell, either. If the genes behaved in such a chaotic manner, doing something different at every moment, there would be little or no coordination between them. They could not possibly regulate cell activity. If a cell was to function at all, there had to be regular patterns in the action of its genes and their production of proteins.

But when Kauffman considered the case where each gene had two inputs, he found an entirely different kind of behavior. The genes would cycle through a relatively limited number of different states. When the cycle ended, it would be repeated all over again. It looked as if this was a kind of orderly activity that could serve as a model for genetic activity in real cells.

It took Kauffman years of work to make these discoveries. At first sight, they do not seem very impressive. After all, it was a very simplified model. Not only had Kauffman assumed that genes could influence each other only by turning each other on and off, he had also assumed that the number of inputs a gene had was always the same. There was no reason that this should be the case. In a real cell, it was more likely that one gene would have no inputs, while another had one, and others three or four.

But Kauffman's discovery was more significant than it might appear. He had discovered that, out of the numerous different kinds of configurations that this complex system could have, there was one that had some very special properties. This implied that something similar might happen in real cells, that there existed a certain kind of genetic input configuration. Furthermore, it looked like a configuration that could arise naturally. If a genetic network did not function in this way, then the cell would die. Kauffman was convinced that he had observed an emergent property that might illuminate a fundamental characteristic of life.

Admittedly, there was no reason that genes in real cells must always have two inputs or even that the average number must be two. But once Kauffman had created his model, he and other scientists were able to introduce further refinements and to consider genetic networks that were more complex. In such cases, the optimum number of inputs might vary, depending on the nature of the system. However, for any given system there was always *some* optimum number that defined the borderline between "frozen" genetic activity and chaos. This has led Kauffman to describe life as something that exists at the edge of chaos. The term "edge of chaos," by the way, is not original with Kauffman. It was coined by the computer scientists Chris Langton and John Holland. However, Kauffman was the first to use it as a description of genetic networks.

Kauffman speculates that all living systems seek out the edge of chaos. It is this fact, he thinks, that makes life and evolution possible. Initially, Kauffman thought that the genes in a living cell would seek out an edge-of-chaos configuration without the intervention of natural selection. This caused him to begin to wonder if natural selection played any role in evolution at all. Perhaps self-organization could account for everything, he speculated. Perhaps the appearance of emergent properties in living systems was the most important factor in the evolution of life.

Nowadays, Kauffman expresses a more conservative viewpoint. Life exists at the edge of chaos, he says "because evolution takes it there." In other words, he believes that natural selection *must* bring life to the edge of chaos if it is to produce viable organisms capable of evolving further.

Kauffman's theory is an elegant one. But is it true? Again, there is no way of knowing. It is impossible to observe all the interactions in a network of genes. Molecular genetics has reached the point where it can identify each gene in a genome. This has been done in only a small number of cases, the bacterium *E. coli,* for example. The human genome is still only incompletely mapped, and no genome of comparable complexity has been mapped, either. Trying to understand the details of all the interactions between real genes would be like trying to understand the complex strategies employed in a game of chess between two grand masters while having only a hazy understanding of the basic rules of the game.

Kauffman has also created computer models that suggest that the effectiveness of natural selection depends on the interconnectedness of genes. His models seem to indicate that if the interconnectedness of genes is not optimal, then one of two things will happen. Either natural selection will not be able to produce optimal adaptations, or mutation will cause adaptations to be lost after they have been painfully gained. He has demonstrated that, *if* real gene networks are interconnected in the same ways as the "genes" in his computer models, then changes in their interconnectedness will have profound effects on evolvability.

Of course the important word here is "if." Since Kauffman's ideas have not been empirically tested, there are questions about the extent to which his models accurately represent biological reality. There is still widespread skepticism among biologists about this matter. Some of them have suggested that one can get anything out of a computer model, depending on the way one sets it up and the numbers one puts in. Many of them feel that input based on observations of real biological organisms is lacking. John Maynard Smith expressed this idea during a debate with Kauffman in 1995. Speaking of the Santa Fe Institute, where Kauffman does his research, Maynard Smith said, "I can spend a whole week there . . . and not hear a single fact." Kauffman responded, "Now that's a fact!" and his remark was greeted with laughter. However, he did not dispute the point that Maynard Smith was making.

Kauffman's work is significant because it suggests that natural selection is not the sole important factor in evolution. According to Kauffman, self-organization must play a role, too. Interestingly, this idea has gained a modest degree of acceptance among some of the people whom Gould has labeled "Darwinian fundamentalists." For example, in *Darwin's*

Dangerous Idea, Daniel Dennett hailed Kauffman as a "meta-engineer" who had added to Darwinian theory by discovering some of the basic rules of design for living organisms and their genetic networks.

However, it will probably be necessary to wait until molecular biology has been developed further before Kauffman's theories can be tested. At the moment there is no way of knowing how significant his ideas will turn out to be. Nevertheless, by demonstrating several plausible ways that self-organization might play a role in evolution, Kauffman has convinced many scientists that it is a phenomenon that cannot be ignored. The genetic networks on which natural selection acts may indeed be complex systems that have emergent (self-organizing) properties.

The Origin of Life

According to Kauffman, life itself may be an emergent property of certain kinds of complex systems, systems that Kauffman calls "autocatalytic." Kauffman's theory is based on the fact that certain biological chemicals, such as proteins, can catalyze the formation of other chemicals. For example, two proteins that I will label A and B may sometimes combine to form protein D. If this reaction proceeds more rapidly when protein C is present, then C is said to be a catalyst. When A and B combine, C remains unchanged. C may speed up the reaction in any of several different ways. For example, C may act as a template that lines A up with B. Proteins have complex three-dimensional shapes, and proteins often fit into one another. If both A and B readily fit into a catalyst protein, they may be joined much more readily than they would be during chance encounters. After it is formed, protein D will simply drift away, allowing C to act as a catalyst once again when it runs into other A and B molecules.

Catalysts often make the reactions that take place in living cells proceed at rates that are 100 million to 100 billion times faster. There are protein catalysts that cause other molecules to bind together and catalysts that act to break them apart. The protein enzymes that act on DNA and RNA are catalysts. Other enzymes allow a cell to make use of food and to regulate its metabolism. If there were no enzyme catalysts, living cells could not exist.

According to Kauffman, it is the action of catalysts that is the foundation of life. His idea can be expressed as follows: Suppose there is one chance in a million that one protein (or other biological chemical) chosen at random will act as a catalyst on a second one that is also chosen at random. If only a small number of proteins are present, then the chances are that there will be no catalytic action. For example, if there are only ten molecules, the chances that any molecule can act as a catalyst is very low. If none can catalyze the formation of any of the others, two molecules may combine with one another on very rare occasions. This combining will be of little significance. The collection of molecules will be nothing but an inert soup.

But, Kauffman notes, if the number of different kind of molecules is increased, the number of possible catalytic reactions will increase much more rapidly than the number of different kinds present. For example, two different kinds of molecules can interact only with themselves, while if there are four different kinds, then the possibilities are much greater. A may combine with itself or with B, C, or D. B may also combine with itself and with any of the other three. The same is true of C and D. Though the number of different kinds has only been doubled, the number of possible reactions has been increased by a factor greater than two.

According to Kauffman's computer models, something very significant happens when a collection of biological chemicals attains a certain degree of complexity. This is best illustrated by an example. If a molecule has a one-in-a-million chance of catalyzing a reaction, and if the system has become complex enough that there are a million possible reactions, then there is a good chance that the molecule will act as a catalyst. There must be a point at which catalysis ceases to be a rare event. The figure of one in a million, by the way, is arbitrary. If the chance is really one in 2 million, then the same thing will happen when there are about 2 million possible reactions. And since the number of reactions increases so rapidly, the number of different kinds of molecules present would not have to be significantly greater.

Kauffman's computer models indicate that the transition to the point at which there is a great deal of catalytic activity is a sudden one. One may have a system in which nothing much seems to be happening. Then the addition of a small number of new kinds of molecules will cause the system to suddenly exhibit a high level of catalytic activity, and it will "spring into life." In Kauffman's terminology, the set has suddenly become auto-catalytic. He believes that the phenomenon of autocatalysis is the foundation of life.

According to Kauffman, such a process could easily have taken place in nature. Originally, there might have been a relatively small number of proteins or other chemicals. But over long periods of time, new kinds of chemicals might have been created by chance processes. This idea is certainly plausible. Scientists believe that various different kinds of amino acids were present on the surface of Earth 3.5 billion years ago. From time to time, they would have combined, forming peptides (small proteins). Under the proper conditions, the peptides could have combined with one another to create even larger molecules. An increase in the variety of different kinds of chemicals present is something that would have happened naturally. If some of these chemicals became enclosed in membranes (and there are plausible ways that this could have happened), they could have been the precursors of the first living cells.

According to Kauffman, autocatalytic sets are capable of mutating and evolving. From time to time, new chemicals may be formed within sets that are already autocatalytic. If they do, the number of possible reactions will increase, and mutation will have taken place. Kauffman asserts that life and the ability to evolve are nothing more than emergent properties that arise when systems of biological chemicals attain a certain degree of complexity.

The theory is appealing, all the more so because it is based on simple concepts. But is it true? No one knows. As I write this, there is no way to test Kauffman's idea experimentally. It is possible to assemble collections of large numbers of different kinds of biological chemicals. A soup of proteins that had 1 or 2 million possible reactions could be assembled in less than a day. But if it were assembled, there would be no way of determining precisely what was going on in the soup. The technology necessary to examine 1 or 2 million different simultaneous reactions does not exist.

Thus we must conclude that Kauffman's idea that life is autocatalytic action is only a hypothesis. I think it is nevertheless a good example of the original new ideas that have arisen in the sciences of complexity. Previously, the scientists who speculated about the origin of life invented some specific scenario or another. None of them were able to show that their scenarios were really more likely than those proposed by other scientists. Kauffman, on the other hand, created a model that said nothing about the precise composition of a set of biological chemicals. On the contrary, he investigated the self-organizing principles that might be at work. By doing so, he was able to produce a hypothesis about the origin of life that was as plausible as any of the others. Furthermore, his theory implies that life created in this way would have automatically been able to evolve. Thus it is a theory not only of the origin of life but also of the evolvability of life.

Punctuated Equilibrium and Complexity

Niles Eldredge and Stephen Jay Gould are not complexity scientists. When they proposed their theory of punctuated equilibrium, neither had ever heard of the sciences of complexity, a field that was only beginning to be developed at that time. Nevertheless, one cannot help but be struck by the parallels between their ideas and those of scientists like Kauffman. Superficially these ideas are quite different. Eldredge and Gould look for patterns at a level above that of the individual organism, while Kauffman speculates about the properties of networks of genes or sets of biological chemicals. Nevertheless, their respective theories have some distinct similarities.

Kauffman looks for patterns that emerge when genes interact with one another. Eldredge and Gould suggest that patterns arise in evolution at another level. While admitting the importance of natural selection, Eldredge and Gould maintain that speciation and species sorting also have roles to play. They see species as real entities, not just as repositories of the genes on which natural selection acts. Furthermore, they see species as entities that interact with one another in complex ways.

The evolutionary "pluralism" that Eldredge and Gould preach betrays an outlook that seems to be much closer to Kauffman's than it is to those of such scientists as Richard Dawkins and John Maynard Smith. Like Kauffman, they attempt to avoid looking at evolution in a purely reductionist manner.

But I think we have to be careful not to make too much of this fact. The question of whether or not a scientific theory can be confirmed is far more important than the way that it is classified. However, if we do look at Eldredge's and Gould's theory as one that embodies ideas about complexity, then it is possible to see one reason why they should be so at odds with such scientists as Dawkins and Maynard Smith. Dawkins and Maynard Smith are reductionists. Eldredge and Gould, like Kauffman, are among those scientists who are beginning to look at evolution in a different way.

Can the difference between the two opposing schools of evolutionary biologists ever be reconciled? I think they probably can. But before this can be done, it will be necessary to gather more evidence about the precise ways in which evolution works. And as we shall soon see, this is exactly what is being done, both in the world of traditional biology and in the world of computer modeling.

Life Inside a Computer

One of the reasons scientists are so interested in the possibility that life might exist elsewhere in our solar system is that, if extraterrestrial life was discovered and studied, then we would gain important insights into the workings of evolution. We would have a better idea as to which characteristics of evolution on Earth are inevitable and which are accidental or dependent on the conditions that exist on our planet. Alien life could differ from terrestrial life in any of a number of different ways. It might make use of an entirely different kind of genetic code, it might embody genetic mechanisms that do not exist here, and it might have evolved in different ways.

We can be reasonably certain that evolution elsewhere would depend on natural selection. No one has ever thought of anything else that could cause organisms to become adapted to their environments. The similarities might end there. But perhaps it would be best not to speculate too much on the ways that extraterrestrial life might be different. There are certain to be surprises that we cannot anticipate. When other forms of life are discovered, the reality might turn out to be stranger than anything conceived by our imaginations.

No real signs of extraterrestrial life have yet been discovered. Scientists have succeeded only in identifying environments that might be hospitable to life, such as the surface of Mars or the ocean that is believed to exist under the icy surface of the Jovian satellite Europa. You may recall that some scientists initially thought they had detected signs of life in a meteorite that originally came from Mars. However, as the Martian meteorite was studied further, scientists reached the consensus that it had probably never contained any life at all.

You might think scientists will probably have to wait a long time, perhaps decades, before they have any alien forms of life to study. However, this is not the case. They are studying alien life now, and they have been for more than a decade. The life they are studying did not come from some other celestial body, however. It is life that lives inside computers. The electronic organisms they have created are born, eat, reproduce, mutate, and die. And because they mutate, natural selection causes them to evolve inside their computer environments.

The pioneer in the field was the biologist Thomas Ray. In 1989 Ray, who was then at the University of Delaware, went to the Santa Fe Institute to talk to scientists there about his idea of creating artificial, evolving life inside a computer. Ray planned to create computer viruslike organisms that differed from ordinary computer viruses in that they would be able to mutate.

Ray realized that such organisms might be dangerous if they escaped from his system and managed to infect other computers. Their ability to evolve might make them especially difficult to eradicate. He therefore planned to run his experiment in a personal computer that had metal bars covering the floppy disk drive and the serial port. This, he believed, would prevent anyone from intentionally or inadvertently transferring them to another computer.

The Santa Fe Institute scientists threw cold water on Ray's ideas. This kind of research program would never work, Ray was told. Computer programs were much too "brittle." If a mutation caused a single instruction to change, then most likely the program would cease to function entirely. Furthermore, the security measures that Ray planned were inadequate. He would have to run the experiment inside a virtual computer.

Ray wasn't a computer scientist; he was a biologist who had picked up a bit of programming knowledge here and there. He had never heard this term before. "What's a virtual computer?" he asked.

A virtual computer, the institute scientists told him, was a simulation of a computer inside a real computer. Creating a virtual computer was something like creating a computer model of an airplane. The only difference was that, in this case, what was being modeled was another computer. If Ray's artificial organisms lived inside the virtual computer, they could never escape because as far as the real computer was concerned, they were nothing but chunks of data. They would behave as computer viruses only inside the virtual computer simulation.

If Ray had been a computer scientist, he might not have succeeded with his artificial life project. Indeed, he might not even have tried. But he was a biologist. Thus he modeled his artificial organisms on real biological organisms, not on existing computer programs. Calling his artificial world Tierra (Spanish for "Earth"), he seeded it with an electronic organism he called the Ancestor. The Ancestor was not a very complicated creature. Where the simplest bacteria have hundreds of genes, the Ancestor had only three; it was only eighty computer instructions long. It was designed to be able to replicate itself, to mutate, and to feed on the energy provided by the computer's CPU (central processing unit). Ray's world also contained a subprogram he called the Reaper. The function of the Reaper was to ensure that all the organisms that were born in Ray's artificial world would die eventually. It was designed to allow the better-adapted organisms to live a little longer and to produce more offspring, but it eliminated them all eventually. This not only prevented overpopulation, it also allowed natural selection to act. If Ray's electronic creatures had been immortal, then all them would have been able to go on passing their genes down to subsequent generations. If natural selection was to be effective, some of the organisms had to be more likely to survive than others.

After the chilly reception of his idea in Santa Fe, Ray really didn't expect much in the way of results. The advice he had been given had convinced him that it might take years before he could observe evolution inside a computer, if he succeeded in doing so at all. But at least he could make some attempts, he decided. Setting up the experiment would give him a better understanding of the problems he would face and might suggest ways in which he could make the idea work.

By early 1990, Ray had his system running. He placed a single Ancestor organism in his artificial world, which consisted of a block of computer memory. The Ancestor promptly replicated itself and replicated again, until the world was about 80 percent full. Then, very quickly, mutant forms began to appear. The first was a creature that was made of some seventy-nine computer instructions, compared with the Ancestor's eighty. It turned out to be better adapted than the replicas of the Ancestor, and it soon began to increase in numbers, while those of the Ancestor diminished. Then new forms began to appear. They were smaller yet. Apparently, with fewer computer instructions, they were able to reproduce more rapidly and thus had an evolutionary edge over their more cumbersome forebears.

Next, even smaller organisms evolved; they were able to reproduce more rapidly yet. Finally, some electronic organisms evolved that were too small to reproduce on their own; they didn't contain the minimum number of computer instructions that would have made reproduction possible. They were parasites, which made use of the reproductive machinery of larger organisms, just as viruses make use of the metabolism of the cells they invade to reproduce themselves. The parasites were only forty-five computer instructions long.

At first the parasites proliferated. But natural selection again came into play. The hosts began to evolve defenses against the parasites, and the latter were nearly driven to extinction. However, evolution continued, and new species of parasites arose, which had the ability to evade the hosts' defenses. An evolutionary arms race was taking place.

It is thought that such arms races are an important part of biological evolution. Predators and prey often evolve together. The prey animals either evolve defenses against their predators or develop traits that make it more likely that they will escape. For example, the members of an antelope species may become swifter in order to better escape lions. The lions will respond by evolving traits that make them better predators. In each species, natural selection will favor traits that make animals better competitors against the other species. Such arms races do not go on forever. For example, as antelopes became faster, there would come a point where they would have to make too many trade-offs to be able to run faster yet. An animal that was all legs and leg muscles would not be well adapted for much but running.

Real evolutionary arms races must go on for many thousands of years, at the very least. But since evolution proceeded so much more rapidly in Ray's digital soup, he was able to observe arms races that played themselves out over periods of minutes rather than millennia. His experiment was an overwhelming success. But Ray did not stop here. He continued his experiments by injecting organisms that had evolved in previous runs into Tierra rather than beginning with the Ancestor. During the course of one series of experiments, he was able to observe 29,000 different types of electronic organisms that appeared at one time or another. These organisms differed widely. For example, Ray saw organisms of about 300 different sizes.

Artificial Life on the Internet

Ray had every reason to be pleased with the success of his experiments. However, he was well aware that Tierra had certain limitations. It was a world that had a limited size, and the electronic creatures that evolved in it did not evolve a great deal of complexity. The organisms in Tierra were single-celled creatures that were much simpler than the simplest bacteria. Tierra did not have different ecological niches as the real world did. The organisms in it fed on "energy" from the computer's CPU, and they all had an equal chance to make use of it. There were no marginal environments, no places where the living was easy, and no environments in which it would have been necessary to evolve different kinds of traits in order to survive.

For these reasons, the phenomena observed in Tierra shed little or no light on the dynamics of biological evolution. Not only were the organisms that inhabited this world too simple, the results were hard to interpret. When Tierra was run on different computers, different kinds of results were obtained, depending on the nature of the computer language that was used. Digital evolution proceeded slowly and gradually in some systems, while something resembling punctuated equilibrium was observed in others. On one system, there was gradual evolution between punctuational events in which speciation occurred. On another system, stasis persisted between the events. It was hard to guess what this meant because there was no theory that allowed scientists to relate the structure of a computer language to the patterns of electronic evolution. Since the patterns of evolution varied from one machine to another, it was hard to make any generalizations.

Ray saw that something could at least be done about the simplicity of the digital organisms and of their environment. He began to think about placing a more complex Ancestor in an environment that was not the same everywhere. Certainly this would give his electronic creatures the opportunity to evolve. There was an obvious way that this could be done. Computers were connected to one another by the Internet. So why not set up an experiment in which digital life could travel from one computer to another? It would then experience a much more diverse environment, and new and surprising results might emerge. There was reason to hope that complex electronic ecologies might develop and that multicellular electronic life would evolve a large variety of different forms.

The computers that harbored the electronic life could go on performing the tasks for which they were ordinarily used. Tierra could be set up as a program that ran in the background, like a screen saver, becoming active only when the computer was idle. The computer could go on performing its usual tasks during the day and make its CPU "food" available to digital organisms when it was idle during the night. Such a scheme had the added advantage that electronic organisms would be encouraged to migrate over the Internet from one site to another. If little or no CPU energy was available in one computer, they would travel to another in a different part of the globe.

Ray named his new experiment Network Tierra. The first experimental runs were performed in 1997 on a network of interlinked computers at a number of different locations around the world, from Japan to Europe. An ancestor containing 640 computer instructions was seeded into the network. It was eight times as large as the original ancestor, which had had only eighty. The original Network Ancestor had ten cells and two cell types. There were eight sensory cells and two reproductive cells containing a total of six genes. The sensory cells were designed to allow organisms to find their way from one Internet site to another.

These multicellular digital organisms were designed to produce daughter organisms that grew from single-celled "embryos" into multicellular creatures in a manner similar to that in which the embryos of real multicellular animals develop. Prior to reproduction, one of the two reproductive cells would die. The other reproductive cell would split in two, producing a single-cell embryo and a reproductive cell that remained part of the original multicellular digital organism. The cell that remained in the organism would then split a second time. This time, both cells were retained, so the "mother" regained her original ten-cell form.

Meanwhile, the daughter cell would split in two, producing a sensory cell and a reproductive cell. The reproductive cell would then split a second time, bringing the total to two. The sensory cell, on the other hand, was genetically programmed to split three times, producing eight cells in the third generation (if the number 1 is doubled three times, the result is 8). When this was accomplished, the daughter had grown into the ten-cell adult form.

After reproducing, the mother organism could either migrate to a different computer on the network or send her daughter to a different computer. The eight-cell sensory system was designed to allow her to choose a suitable location. All locations were not equal. A faster computer, for example, would produce more of the CPU energy on which Ray's creatures thrived.

There was a Reaper at each computer node. When the number of cells on a node (i.e. on one individual computer) reached a preset maximum, the Reaper went into action, killing one organism for each organism that was born on or migrated to that node. Finally, Ray programmed periodic "apocalypses" in which *all* the organisms on a node were killed off. Since nonmigrating organisms would eventually be killed in an apocalypse, they would be eliminated by natural selection. Only organisms that periodically traveled from one node to another would survive.

In the initial runs, three different Ancestors were injected into the network. Almost immediately, a problem developed. Ray's digital organisms showed a tendency to lose their sensory cells while multiplying the number of reproductive cells. Soon, organisms developed which were composed of 256 reproductive cells (256 was the maximum that the experiment allowed) and nothing else. These organisms avoided being wiped out by apocalypses by migrating randomly. Since they made no attempt to choose one node over another, sensory cells were not needed. Meanwhile, the large number of reproductive cells ensured that they would have numerous offspring and thus attain evolutionary success.

This was evolution, but it was not the kind of evolution Ray was interested in. One of the purposes of the network experiment was to see if the digital organisms would evolve new cell types. He wasn't interested in evolving creatures that would breed and do little else. So he tinkered with the design of the Ancestors, trying to see if he could produce organisms that found the retention of sensory cells to be necessary. He soon achieved success, producing organisms that retained their sensory apparatus indefinitely, during long bouts of evolution.

The organisms didn't produce any new cell types, but they did evolve into new species. When digital species evolved, they were usually not as large as the 256-cell breeder had been; most had between four and thirty-two cells. But they did exhibit a satisfying degree of diversity. For example, there was a great deal of variety in the ways Ray's organisms foraged for resources, primarily the CPU "energy" on which they lived.

One of the objectives of Ray's experiment, the evolution of new cell types, has not been achieved. However, an important evolutionary phenomenon, gene duplication, has appeared. Some of Ray's digital organisms have evolved duplicate copies of some of their genes. Once this gene duplication occurs, the two initially identical genes have a tendency to evolve in different directions, causing the total number of distinct genes to increase. Gene duplication is believed to have played an important role in biological evolution, especially in the evolution of multicellular life. If more complex organisms are to evolve, the number of different kinds of genes must increase. A bacterium can get by with a few hundred or a few thousand genes, but such a small genome would not be sufficient for producing a multicellular organism. For example, it is no accident that we have nearly a hundred times as many genes as the *E. coli* bacteria that live in our digestive tracts.

Evolvability

The network Tierra experiment shows a great deal of promise. However, Ray suspects that before it can proceed very far, complexity scientists must learn to better understand the phenomenon of evolvability. Understanding evolvability is especially important in the case of digital evolution since there are two different kinds of factors that affect evolution. One kind is like those that affect the evolution of biological organisms; the other is unique to a particular computer environment. The nature of the computer language used, for example, seems to be quite important. And of course no one yet knows why.

Ray notes that there has been a lot of work done in certain areas of the science of complexity that might shed light on some of these questions. This work deals with the manner in which an evolving system becomes well adapted. But, as Ray points out, there are other factors to be considered too. Biological evolution has had a number of major transitions, such as the evolution of eukaryotic cells and the evolution of multicellular life. Ray would like to discover what conditions might allow analogous major transitions to appear in digital evolution. One example of this would be the evolution of new cell types.

Ray's studies of evolvability are only beginning. Before writing this chapter, I contacted him and inquired about the current state of his research. He replied that he was in the process of preparing a paper in which he would attempt to suggest how some of the questions concerning digital evolvability might be approached. Obviously there is a long way to go.

Digital Evolution vs. Biological Evolution

Earlier in this chapter I discussed some of the work of Stuart Kauffman, who, among other things, has created computer models of gene networks. I think it would be illuminating to compare his approach to evolution with Ray's. There is one respect in which the two approaches are very different.

Kauffman has created computer models of real biological systems. Ray does the opposite. His artificial life experiments are not intended to be models of biological evolution. Ray is not trying to understand how biological organisms evolve or the history of life on Earth. On the contrary, he uses his knowledge of these organisms to set up a different kind of evolution inside a computer. He seeks to understand how evolution works, whatever it is that might be evolving.

The world in which Ray's organisms live is very much unlike the world we inhabit. It has no geometry, for example. Some artificial life systems have been developed in which digital organisms live on two-dimensional grids. This feature has not been incorporated into Tierra or Network Tierra. Ray's electronic organisms cannot move around inside a computer node. They simply occupy a chunk of computer memory and feed on energy provided by the CPU. Since they can travel from one computer to another almost instantaneously, there are no "distances" in their world. There are simply a number of different locations to which they can migrate.

For these and for other reasons, it is likely that evolution will proceed differently in Ray's electronic world than it does in ours. If it does, the differences are likely to be as illuminating as similarities would be, especially if Ray discovers general laws about evolution or about evolvability. He could conceivably discover that any evolving system will exhibit certain kinds of emergent properties.

Evolution and the Sciences of Complexity

I have discussed only a small part of the work being done in the sciences of complexity that might have implications for our understanding of biological evolution. But, by discussing two different approaches in detail, I hope I have given you an idea of the wide variety of work that is being done in the sciences of complexity and of the kinds of results that might be attained in the future.

Some of the work that has been done in the field has generated a certain amount of controversy. Many complexity scientists view biological structure and evolution as complex systems that have emergent properties. This causes them to suspect that natural selection is not all there is to evolution. But they have not yet conclusively demonstrated this, and many orthodox biologists discount the importance of the research that they have done. These biologists often doubt that computer models can really be accurate representations of something as complex as a genome or an organism or evolutionary patterns. This controversy is not likely to be settled soon. The sciences of complexity is a relatively new field, one that has not yet fulfilled its promise. The findings obtained so far sometimes seem illuminating, but it is still possible to doubt that they reflect biological reality.

The controversy is not likely to be settled until more work is done both in the sciences of complexity and in the biological sciences. If some of the complexity scientists' results—those of Stuart Kauffman, for example—are at some point tested in a biological laboratory, a significant step forward will have been taken. Similarly, if Thomas Ray discovers universal principles of evolvability that seem to describe both digital and biological evolution, a connection between the two fields will have been established.

Alternatively, further research may show that some of the results obtained by complexity scientists are not as relevant as they initially seemed to be. If this happens, however, I don't think that the work that produced these results will have been done in vain. Science learns from its mistakes. When the wrong road is occasionally taken, this often leads to a better understanding of the problems being faced and may suggest approaches that will work.

There does seem to be a similarity between the ideas of the complexity scientists and those of Eldredge and Gould, who understand evolution as something that depends on factors that operate at different levels of complexity, as something that cannot be completely explained by a reductionist approach that considers only natural selection and a few subsidiary principles, such as sex (sex causes gene pools to become more diverse by mixing up the genomes of the parents when offspring are produced). However, at the moment, the sciences of complexity have not yet produced the kinds of results that would confirm or contradict Eldredge's and Gould's ideas.

The sciences of complexity is a field full of promise. But only time can tell how much of this promise will be fulfilled. Only time will tell how much insight complexity research will shed on the mechanisms of evolution.

Evolutionary Psychology

In most animal species, including *Homo sapiens,* the males court the females in any of a variety of different ways. In other species, they fight to obtain mates. For example, a peacock displays his brightly colored tail plumage to attract females. The males of the African village weaverbird species court females by inviting them to inspect nests they have built. Scorpionfly males attempt to induce females to mate with them by offering nuptial gifts, usually dead insects. Elephant seals, on the other hand, don't go in for elaborate mating rituals; they engage in combat, often fighting until one of the competing seals lies exhausted on the beach.

And of course these strategies generally work. Female elephant seals don't mate with their defeated suitors; they become members of the harems of the dominant males. Female peacocks are attracted by displays of bright plumage, and weaverbird females consent to mate with the males who have built the best nests. In general, the females are much more choosy than the males. In fact, in some species the males are absurdly undiscriminating. A replica of a female turkey's head that has been suspended some appropriate distance above the ground is sufficient to induce a male turkey to engage in ritual mating displays. The male may even attempt to mate with the bodiless female.

It is not difficult to find plausible reasons why females should exercise choice and why the males of many species should be so willing to mate with any available female. In fact, the difference between male and female mating behaviors can be explained by a theory that is so plausible that it is difficult to imagine that it isn't true.

In 1972 the Rutgers University biologist Robert Trivers (who was then an untenured professor at Harvard) introduced the idea of "parental investment," which he defined as "any investment by the parent in an individual offspring that increases the offspring's chance of surviving (and hence reproductive success) at the cost of the parent's ability to invest in other offspring." Parental investment included the costs of producing an egg or a sperm, either incubating fertilized eggs or undergoing pregnancy, and then rearing the offspring. It is obvious, Trivers noted, that the parental investments of males and females are not equal. Eggs are larger than sperm, and more resources are required to produce them. Additional biological costs are incurred during incubation or pregnancy. And, more often than not, it is the female who invests more resources in an offspring after birth.

According to Trivers, this simple fact makes it possible to understand such things are male eagerness and female coyness, many aspects of courtship, the rearing of offspring and such phenomena as fidelity and infidelity. Because the costs to a male parent are usually low, it is to the advantage of the males to spread their sperm among as many different partners as possible. Females invest more heavily in their offspring. Thus it is to their advantage to select genetically superior males and to repel the advances of those that do not measure up. If they select males with superior genes, their offspring will have better genetic endowments, and the continued survival of the genes that are present in the female will be more likely.

Naturally Trivers was not suggesting that the females make conscious decisions. In most species they can't. But of course they don't have to. If selfish genes (and remember that the concept of selfish genes is a part of orthodox Darwinian theory) are able to cause an organism to behave in a way that ensures the genes' survival in future generations, then those genes will proliferate. Males are relatively undiscriminating because it is in their genetic interest to spread their genes as widely as possible. Because the females incur greater costs, they cannot do this. If they have only one clutch of eggs or one litter of offspring per year, mating with more than one male will produce no benefit. They are better off choosing a single male who confirms his genetic worth by winning a combat or by showing that he was as fit and healthy enough to have bright tail plumage or by demonstrating the ability to build a superior nest.

If a theory is plausible, it does not necessarily follow that it is true. Theories must be tested. Now at first glance, there doesn't seem to be any obvious way to subject Trivers's theory to empirical confirmation. But a little further thought shows that this task is really not so hopeless. If the parental investment theory is a correct description of mating behavior, then members of species in which the male parental investment is unusually large should behave somewhat differently. Indeed, this is just what is observed. For example, in the pipefish species, it is the male who cares for the eggs. He places them in a pouch and nourishes them with his bloodstream. As expected, female pipefish display a very different kind of behavior than the females of most other species. They are the ones who play the "masculine" role, seeking out males and performing the mating rituals. Similarly, among the bird species known as phalaropes, it is the males who sit on the eggs, giving the females the opportunity to leave and seek other mates. In most bird species, it is the males who are larger and more colorful. But among the plalaropes the females are bigger and display brighter plumage. Furthermore, similar role reversals are seen among other species in which male parental investment is large, including some birds, the Panamanian poison-arrow frog, a species of water bug, and the Mormon cricket.

Human Mating Behavior

But what about the human species? Do we exhibit types of mating behavior that can be attributed to our evolutionary history? Does our genetic endowment cause men to be naturally somewhat promiscuous and women coy? Do our genes play a role in our selection of sexual and marriage partners?

During most of the twentieth century, the traditional answer to these and similar questions was "no." Human beings differed from other animals, the anthropologists and other social scientists said, because only they had a culture. The various societies that existed in the world socialized young human beings, who generally accepted the cultural values those societies imposed—monogamy or polygamy, for example. Cultural values caused virginity to be more highly valued in some societies than in others. Culture made us what we were in these and in numerous different ways. Our biological heritage, to be sure, was important. It provided us with sex drives, for example. However, culture determined the ways in which our sexuality was expressed, while molding "human nature" in numerous other ways as well.

Some of the anthropological studies that were carried out seemed to confirm this view. Of these, the most famous and most influential was one done by anthropologist Margaret Mead, reported in her book *Coming of*

Age in Samoa, published in 1928. According to Mead, Samoan culture lacked many of the characteristics that were common in the West. There were no status hierarchies, for example, There was no tension about sex and no adolescent turmoil. Girls postponed marriage "through as many years of casual love-making as possible." The Samoans experienced no feelings of jealousy. Cuckolded husbands exhibited no anger, for example. The message was clear. Here was a culture that was wholly unlike Western culture of the early twentieth century. Obviously, human beings were infinitely malleable.

Mead's book was followed by another, *Sex and Temperament in Three Primitive Societies,* published in 1935. In this book, Mead reported on her studies of three preliterate societies in New Guinea, known as the Arapesh, the Mundugumor, and the Tchambuli. She found that the Arapesh were a gentle, unassertive, cooperative people. Interactions among the Mundugumor, on the other hand, were characterized by suspicion and hostility. And in the third society, that of the Tchambuli, gender roles were practically the reverse of those that were common in Western culture at the time. For example, not only were the women believed to be more highly sexed than the men, they were more dominant and sexually aggressive, the men were naturally passive, responsive, and interested in children.

The evidence seemed conclusive. These findings did not correspond very well to common Western folk wisdom about human sexual behavior. According to this folk wisdom, men were the natural sexual aggressors and more undiscriminating about sexual partners than women were. They were the ones who sought women's sexual favors, while the women were more coy and selective.

But perhaps this folk wisdom was only a reflection of Western values. If Mead and other social scientists were right, we were as influenced by our culture as the people Mead and other anthropologists studied. Indeed, the scientific evidence seemed to indicate that this was the case. To be sure, western folk wisdom painted a picture of human sexual behavior that seemed to be very much like the one scientists observed in many species of animals. But was this really so meaningful? Perhaps scientists were looking at the behavior of these animals through anthropomorphic eyes. In any case, if animals behaved in a certain way, it did not necessarily follow that humans must too.

Sexual and Mating Preferences

The view promulgated by the anthropologists was widely accepted. But it has not prevented psychologists from going ahead and studying sexual and mating behavior anyway. The results they have obtained are revealing. They have found that men commonly prefer partners who are younger than they are, and that when they are looking for sexual partners, they are markedly less discriminating than women. In some cases the differences are striking. For example, in one psychological study, both men and women were approached by members of the opposite sex whom they did not know. When the strangers expressed a desire to have sex, some 75 percent of the men agreed, while none of the women did. And it is not even necessary to carry out very comprehensive studies in order to glean certain other important facts. For example, it is men who employ the services of prostitutes and who are the largest consumers of pornography. And it is men who more frequently find themselves desirous of extramarital affairs.

According to evolutionary psychologist David Buss, certain patterns can be observed in male and female sexual behavior and preferences in numerous societies in different parts of the world. In 1989 Buss published a study of mate preferences in some thirty-seven different cultures. Buss and his collaborators surveyed some 10,047 individuals on six continents and five islands. The participants came from polygamous societies, such as those in Nigeria and Zambia, as well as from cultures that were more monogamous. Scandinavian countries, in which living together was very common, were included in the study, as well as nations like Greece and Bulgaria where this practice is frowned on. Buss found that in all the societies men placed a higher value on youth and physical attractiveness in potential mates, while women wanted men with economic resources. In none of the societies were postmenopausal women considered to be very attractive, while men of the same age often did not experience difficulty attracting mates. And of course there were other differences, as well. For example, women preferred men who were or greater than average height, and they rated dependability as a desired quality more often than men did.

It should be noted that in this particular study, Buss was looking at preferences concerning potential marriage partners, not about casual sex partners. These two sets of preferences are not necessarily the same. In fact, there is some evidence that they are different. Men—American college men, at least—seem to place a higher value on intelligence in potential mates than in short-term partners; they seek out casual sex partners who seem to have greater sex drive and sexual experience than they would prefer in a wife. But whatever it is that is being studied, whether standards of attractiveness, qualities desired in a short-term partner, or qualities sought in a mate, there are always differences between men and women. Furthermore, some of the differences appear to be universal; they are found in all the cultures that Buss and his colleagues studied.

But not all social scientists are convinced. There have been numerous criticisms of Buss's results. For example, it has been pointed out that a great majority of the different cultures he studied were European or European-influenced and had urbanized cash economies. The similarity of the responses of people in different cultures might reflect only the fact that these people were subject to similar cultural influences. In other words, according to the critics, the responses might have had nothing to do with natural selection or genetics. One can make an analogy by pointing out that Western cultural influences cause business suits to be worn in most of these cultures. Yet no one would ascribe the wearing of business suits to genetic causes. Another objection is based on the observation that what people say and what they do are often quite different. Nearly all men will say that they prefer beautiful women, but most do not marry women who are considered to be beautiful.

Furthermore, people's responses to questions about their sexual behavior may not be true. As the British psychologist Dorothy Einon points out, there is a discrepancy between the number of sexual partners claimed by men and by women. The former figures are typically three or four times higher than the latter. The difference can be explained in two ways. Either there are enough sexually hyperactive women (such as prostitutes) to make up the difference, or people do not tell the truth. According to studies that Einon carried out on British and French populations, the number of sexually hyperactive women would have to be extraordinarily high to account for the gap. Einon concludes that the discrepancy can be better explained by concluding that men have a tendency to boast, while women often give overly modest answers. (But of course, if this is true, it would in itself be a sexual difference.)

Anthropologist Mildred Dickemann has been especially critical of Buss's studies. According to Dickemann, the majority of anthropological studies indicate that sexual behavior and practices vary widely across cultures. The existence of such differences, Dickemann says, contradicts Buss's assumption "that human reproductive behavior is a set of invariant responses arising from some set of invariant evolutionary dicta."

Buss's findings are somewhat controversial. However, many psychologists believe that, in spite of defects in the study, the results are still valid. They doubt the conclusion that culture molds behavior as much as anthropologists and other social scientists think it does. One of the reasons they doubt it is the fact that the validity of some of the work of Margaret Mead has been called into question.

The work of Margaret Mead was enormously influential in establishing the idea that cultural influences are all-pervasive. But in 1983, questions were raised about the accuracy of the anthropological study she performed in Samoa. In that year anthropologist Derek Freeman published a book titled *Margaret Mead and Samoa: The Making and Unmaking of an Anthropological Myth*. Freeman pointed out that Mead had spent only nine months in Samoa and that she hadn't spoken the Samoan language when she got there. She hadn't lived among the Samoans and had depended on interviews with informants to learn about the culture. Freeman, who lived in Samoa for almost six years, claimed that Mead's description of Samoan society had been misleading and inaccurate. The reality, he said, was very different from the picture Mead had painted in a number of important respects. Freeman found that, contrary to what Mead had claimed, virginity was highly valued in Samoan society. He found that strong sexual jealousy did exist. What Mead had described as a tension-free society was actually one in which rape and reprisals by the rape victim's family were frequent. Mead had said that there was no adolescent turmoil, but Freeman and other anthropologists found that there was widespread adolescent resentment and delinquency. Mead's findings, he concluded, just didn't stand up.

It looked as though considerable doubt had been thrown not only on Mead's conclusions about Samoan society but also on the idea of cultural relativism in general. After all, Mead's Samoan study had been influential in establishing this view. But Freeman's findings aren't quite as conclusive as they might appear to be. Freeman studied a different village than the one studied by Mead, and he did his work some four decades later. In the intervening years, a U.S. military base had been established in Samoa, which may have had an influence on the behavior of Samoans who worked at the base and on Samoan culture in general. Freeman interviewed some of Mead's informants in the role of an honorary chieftain. If they told him a different story, this could have partly been due to the fact that an elderly woman might tell a different story to a high-ranking male than the one she had told to a young woman when she was an adolescent. Some psychologists have described Mead as the victim of a hoax. However, the facts are not perfectly clear.

It has been claimed that Mead's assessments of other societies were not so accurate, either. However, this is also a matter of controversy. Questions have been asked about the validity of Mead's work, but it would be going too far to say that her ideas have been discredited. It appears that all we can really say is that Mead was committed to a theory of anthropological relativism. She believed that human behavior was molded by culture, and she found evidence to support this view. Scientists who do not hold the same view have found reasons to call some of her findings into question.

Mead was one of the most influential anthropologists of the twentieth century. Her work had seemed to bolster the view that culture, not genetics, was the main determinant of the thing we call "human nature." Although it would be difficult to reach any final verdict, there seems to be some reason to entertain doubt about the validity of some of Mead's work. As a result, the doctrine of cultural relativism seems less firmly established than it was a decade or two ago. This, in turn, bolsters the idea that studies such as the one carried out by Buss have demonstrated real differences between male and female behavior. It is true that Buss's studies have been criticized, so perhaps they cannot be considered to be entirely conclusive. Nevertheless, Buss's studies do show that looking for kinds of human behavior that might have a genetic basis can at least be a fruitful line of research.

The Standard Social Science Model

In recent years, two University of California evolutionary psychologists, Leda Cosmides and John Tooby (Tooby was trained as an anthropologist), have been attempting to show that many human behavior traits have a genetic foundation. They call the work they do "evolutionary psychology," and they claim that research that they and other scientists have performed have cast doubt on what they call the Standard Social Science Model, or SSSM.

SSSM is Cosmides' and Tooby's term for what I have been calling the doctrine of cultural relativism. They prefer to call it a model to emphasize the fact that other models of human behavior are possible. According to Cosmides and Tooby, the SSSM has reigned for a century. During that time, they say, it has provided a framework for psychology and the other social sciences. According to the SSSM, Cosmides and Tooby go on, human genetic endowment cannot explain the diversity of human cultures. Social scientists have therefore concluded that human genetic endowment "cannot be the cause of the mental organization of adult humans, their social systems, their culture, historical change, and so on."

Cosmides and Tooby think that this view is incorrect. They believe that it is possible to find certain human traits that are universal. If such traits are found, they say, then it is possible to conclude that they must have genetic causes, and it should be possible to find evolutionary explanations for them. Natural selection, after all, designed the human brain, just as it designed the human hand and our upright posture. If some human beings, or human ancestors, had mental traits that made it more likely that they would survive and reproduce, then the genes that produced those traits would spread through human (or prehuman) populations.

The characteristics of human mating behavior that David Buss claimed to have found would be examples of such universal human traits. According to Buss, men prefer attractive younger women in all human cultures, and most of the men in any culture do not find postmenopausal women to be especially attractive. It isn't difficult to see how this trait might have been molded by natural selection. It is having offspring that passes an individual's genes on to succeeding generations. If an individual has genes that cause a greater number of offspring, then natural selection should make those genes more likely to spread throughout a population. A man who has a vigorous, healthy mate is likely to pass on more of his genes. She is more likely to give birth to healthy offspring, and she will be better able to care for them after they are born. If a man were attracted to women exhibiting signs of poor health, there would be a much greater chance that his line would die out.

During most of our evolutionary history there must have been a strong correlation between attractiveness and health. When our ancestors lived on the African savannas, disease and infection with parasites must have been quite common. Obviously this is no longer true today for many of us, but modern civilization has existed for a very short time compared to the 2 million years that have elapsed since the first members of the genus *Homo* evolved, and to the 100,000 years or so that have gone by since modern *Homo sapiens* appeared. If the ideas of the evolutionary psychologists about evolved human behavioral traits are correct, then those traits must have evolved to adapt us to conditions that existed long ago.

Some evidence seems to support the idea that human standards of physical attractiveness were well correlated with health in ancestral times. Psychological studies have shown that we generally find members of the opposite sex to be most attractive when their features are symmetrical. This is significant because injuries and infection with parasites produce asymmetries in the human body. Human features also become more asymmetrical with advancing age. Furthermore, the preference for symmetry seems to exist in all the different cultures that have been studied, while other standards of physical beauty vary. Some cultures place special emphasis on certain physical features, such as eyes, ears, or genitals. In some cultures a slim body build is considered attractive, while in others plumpness is admired. Preferences concerning lightness or darkness of skin color and breast size and shape also vary. It appears that human males prefer symmetrical features for the same reason that peahens prefer peacocks with brightly colored tails. Both are indications of health and freedom from parasites.

Mental Modules

Finding evolutionary explanations for human behavioral traits is only part of evolutionary psychology. Evolutionary psychology is based on a particular model of the human mind and human behavior—one that differs from the SSSM in important respects. This model is capable of suggesting some very interesting lines of research.

Leda Cosmides and John Tooby compare the human mind to a computer—not to a general-purpose computer but to one that is made up of a number of different modules that perform specialized tasks. These modules exist because natural selection has created them to enable human beings to perform different adaptive tasks. Or, as Cosmides and Tooby put it, "Our neural circuits were designed by natural selection to solve problems that our ancestors faced during our species' evolutionary history."

The reason we have one set of circuits, Cosmides and Tooby say, is that sets of circuits were "better at solving problems that our ancestors faced during our species' evolutionary history than alternative circuits were." Consequently we have neural circuits in our brains that enable us to learn language and recognize faces, for analyzing the shapes of objects, for judging distances, for selecting mates, for seeking foods that will remedy nutritional deficiencies, and for doing countless other things. These mental circuits, Cosmides and Tooby say, did not evolve to adapt us to modern environments. We have spent over 99 percent of our evolutionary history living in hunting-and-gathering societies. During a period of time that may have been as long as 10 million years, natural selection slowly modified our ancestors' brains, adapting them to life on the African savannas.

Thus evolutionary psychology does not try to explain human behavior patterns by showing that they make us more fit under modern conditions. Agriculture was invented only about 10,000 years ago, and civilization is an even more recent phenomenon. It is unlikely that natural selection has modified our brains to any noticeable degree in so short a period of time.

According to Cosmides and Tooby, the principles of evolutionary psychology can be applied to any psychological topic, "including, sex and sexuality, how and why people cooperate, whether people are rational, how babies see the world, conformity, aggression, hearing, eating, hypnosis, schizophrenia and on and on." They say that anyone who is attempting to understand human behavior should ask the following questions:

1. Where in the brain are the relevant circuits and how, physically, do they work?

2. What kind of information is being processed by these circuits?

3. What information-processing programs do these circuits embody?

4. What were these circuits designed to accomplish (in a hunter–gatherer context)?

In other words, if we want to understand why human beings behave the way they do, we should look for specialized brain modules that induce them to do things and then try to understand why such behavior would have been advantageous in a hunter–gatherer society.

The studies of human sexual partner and mate preferences are examples of this kind of approach. The evolutionary psychologists who have studied human sexual and mating behaviors have looked for specialized circuits in the brain (for example, a mate preference module) and have tried to explain why it would be useful to have such circuits in a hunting-and-gathering society. You may recall, for example, that they found that males seemed to prefer females who had physical attributes that could be associated with good health. Males who had this preference would have had more offspring, causing the trait to be preserved by natural selection. Males who did not have this kind of preference would have fewer offspring, causing their genes to eventually be eliminated from ancestral populations.

Cooperation

Human beings cooperate with and help one another. But why? If our behavior is influenced by selfish genes, why do we not go around looking out only for our own interests? Why should we engage in behavior that only benefits other individuals? At first sight, it appears that human beings are behaving oddly when they rescue strangers from drowning, or give money to the homeless, or do any of a large number of other things that they are known to do.

Many people, including some professional anthropologists, believe that cooperative and altruistic behaviors evolved because they promote group cohesion or benefit human groups (such as hunting–gathering bands) in some other way. But this idea cannot be correct. Natural selection produces traits that benefit the *individual*. Traits that do not benefit the individual rapidly disappear. For example, suppose that there is a group of inbreeding hunter–gatherers in which most of the members altruistically share food. Suppose that they do so because their genetic makeup induces them to do so. Suppose the group contains one or a small number of individuals who behave selfishly by accepting all the food that is offered to them while keeping for themselves all the food they obtain. Then it will be the selfish individuals who are more likely to survive and produce offspring. In time, the genes that cause them to be selfish will spread through the community. Selfish individuals will then predominate, and the group as a whole will be worse off than it was before.

Some kinds of altruistic behavior do benefit the individual. A mother who cares for her offspring is improving the chance that her genes will survive in future generations. My brother and I share 50 percent of our genes. If I help him, thus making it more likely that he will survive and reproduce, more of my genes will be propagated into future generations also. My selfish genes, in other words, should make me want to behave in such a way that my close relatives will benefit. But they should not make me be especially helpful to individuals who are not close relatives, especially if I incur some cost by doing so.

As a matter of fact, this is precisely what we see in nature. In the great majority of animal species, individuals do not behave altruistically or cooperate with one another. They compete. To be sure, there are some exceptions.

For example, vampire bats sometimes fail to find a suitable large animal from which they can obtain a meal or are prevented from drinking as much blood as they would like. As a result they sometimes return to their roosts hungry. Failing to obtain a meal is a serious matter. If a vampire bat goes sixty hours without obtaining any blood, it can easily starve to death.

But this rarely happens. If another bat has drunk more blood than it needs, it will regurgitate some of the blood in order to feed another, hungry bat. It obtains no benefit from doing so. However, since the other bats exhibit the same behavior, it is likely to be given some regurgitated blood some night when it is hungry. The bats cooperate by sharing food. By doing so, they benefit one another as individuals. The bats that feed one another may not be closely related. But since vampire bats roost in the same place for long periods, they get to know one another as individuals. Thus they are able to return one another's generosity. A bat that has donated a meal in the past will probably get one at some time or another from the bat it aided. And a bat that has refused blood is likely to be refused blood in return. Vampire bats seem to be very good at keeping score.

Humans tend to be good at keeping score, too. This is the basis of the "you scratch my back, I'll scratch yours" principle. It seems likely that when they engage in behavior that is helpful to other human beings, they are behaving like vampire bats. To be sure, few of us expect any return favors from the homeless person to whom we give money on the street. However, in most cases we do demand something. If I invite you to a dinner party, it isn't unreasonable of me to expect that the invitation will be returned at some point. If we are close friends who have given one another Christmas presents or sent cards to one another in the past, I am likely to be puzzled or angry if, some year, I fail to receive one from you. If you do a favor for someone you work with, you will probably expect to obtain a return favor from that person sometime in the future. You may not consciously be thinking of this when you help your coworker, but you are likely to be puzzled or miffed if that person seems unwilling to give you a little help when you need it.

In this discussion of human reciprocity, I have done little but appeal to folk wisdom about the ways in which human beings behave and refer to customs with which we are all familiar. In other words, there is nothing very "scientific" about what I have said. But there are ways of making such a discussion scientific. One of them is to apply the principles of evolutionary psychology.

If these principles are correct, then our brains ought to contain neural circuits designed to make us behave in the appropriate manner toward people who cooperate with us and people who cheat by not returning favors. This idea may seem a little far fetched to you at first. After all, we are surely intelligent enough to tell who is behaving toward us in a friendly manner and who isn't. Why would we need to have some kind of special module in the brain for that purpose?

If you think about it a little, you may see that matters aren't that simple. Remember the hypothetical example of the tribe of altruists that contained a few selfish members? One of the reasons the selfish individuals would have been more likely to survive and spread their genes through the population was that their behavior was not resented. Sharing doesn't work if we don't have a predilection for treating cheaters appropriately, for example, by refusing to share with them. We need to have something like the circuits in the brains of the vampire bats, who will refuse to regurgitate blood for the bats who have refused them in the past. If sharing is advantageous, then natural selection should cause us to desire to share. And we also need something that will make us treat cheaters appropriately; natural selection should provide that, too. Or at least this is how evolutionary psychologists like Cosmides and Tooby view matters.

But how can we tell whether or not this theory is correct? There is one way to do that: by performing experiments that will test the hypothesis.

The Wason Selection Task

Suppose you are shown the four cards marked with the following symbols:

D F 3 7

You are then asked which two cards you must turn over to see if any of the cards violate the following rule:

If the letter D is on one side,
then there will be a numeral 3 on the other.

Which two cards do you turn over? More than 75 percent of the people who take this test give the wrong answer. The correct answer is that one must turn over the card with the D and the one with the 7. If the D card does not have a 3 on the other side, the rule is violated. If the 7 card *does* have a D on the opposite side, the rule is also violated.

This experiment is called the Wason selection task after the psychologist Peter Wason, who invented it. Wason devised the task to see how logical human thinking was. He was inspired by an idea of the Anglo-Austrian philosopher of science Karl Popper. In his book *The Logic of Scientific*

Discovery, Popper had maintained that no scientific theory could be conclusively and finally "proved." A theory, Popper said, was accepted if repeated attempts to falsify the theory failed. Wason's card test was devised to see if that was the way people actually thought. When they were asked which cards they would pick, they were being asked how they would go about trying to falsify a rule.

Wason's results seemed to indicate that the majority of people had no clue as to how they should go about falsifying a hypothesis. They habitually thought in unscientific, illogical ways. Even most of the college students who had taken courses in logic gave the wrong answer.

But then Leda Cosmides discovered something very interesting about the test. She found that if the turning over of cards was dispensed with and the test was modified in such a way as to put it in a familiar social context, then people did get the right answer more often than not. For example, if people are asked the following question: "If you are a bouncer in a bar, and if only people eighteen or older are allowed to drink beer, which do you have to check: a beer drinker, a Coke drinker, a sixteen-year-old and an eighteen-year-old?" Most people correctly answer that the bouncer must make sure that the beer drinker is eighteen or older and he must make sure that the sixteen-year-old is not drinking beer. This test is logically the same as Wason's. Yet people's success rates are quite different.

If the social context is somewhat more exotic, people still do well. For example, in one experiment, devised by Cosmides and Tooby, the subjects were asked to imagine that there existed a powerful chief named Big Kiku on an island in the Pacific. Big Kiku had a habit of demanding that his followers tattoo their faces. One night, four hungry men of another tribe stumbled on his camp and asked to be fed with cassava root. Big Kiku replied that if they tattooed their faces, then they would be fed in the morning. Now suppose Big Kiku tells you sometime later that the first man got a tattoo, while the second was given nothing to eat. The third did not get a tattoo, and the fourth received a large cassava root. Which of the men must you inquire about further if you want to determine whether or not Big Kiku kept his end of the bargain? About 75 percent of the people who are given this problem correctly answer that they must ask Big Kiku whether the first man (who got the tattoo) was fed and whether the second (who was sent away hungry) got a tattoo. The other two are not relevant because Big Kiku would not have broken his promise if he refused to feed the man without a tattoo, or if the fourth man—who received a cassava root—had not been tattooed. Big Kiku could have fed the fourth on whim without breaking any promises.

Cosmides, Tooby, and other evolutionary psychologists continued their Wason selection test experiments for eight years. They devised tests of a large variety of different kinds. After gathering together all the results, Cosmides and Tooby wrote a paper in which they claimed that they had shown that the human brain had a "cheater detection" mechanism that had evolved for the purpose of determining whether bargains and social contracts had been adhered to. "These findings," they concluded, "strongly support the hypothesis that the human mind includes cognitive procedures that are adaptations for reasoning about social exchange."

According to Cosmides and Tooby, the experiments undermined the idea that ideas about social exchange were culturally transmitted. The cheater detection mechanism looked like something that was innate in the human mind, something that had evolved. People made implicit assumptions about social contracts. This was why they found Wason selection task problems so much easier to deal with when they were placed in social contexts which involved the making of a bargain or the detection of cheating. They admitted that there were many questions that had not been answered. For example: When would human beings cooperate with one another on a long-term rather than a short-term basis? If the resources exchanged are easier or harder to obtain, how will this affect the ways in which they are shared? What role is played by groups and coalitions in shaping patterns of assistance? What roles are played by aggression, retaliation, and status? However, they did not doubt that they had opened up a line of research that promised to allow evolutionary psychologists to learn a great deal about the evolved mechanisms in the human mind.

Evolutionary psychologists have studied a variety of different kinds of human behavior besides mating patterns and sharing. The topics they have studied include parental care, "morning sickness" during pregnancy, the nature of play fighting, perception and language, evolved responses to certain types of landscapes, and rape.

The last topic evoked a great deal of controversy in 2000 when biologist Randy Thornhill and Anthropologist Craig T. Palmer published a book entitled *A Natural History of Rape* (MIT Press, 2000) in which they made the claim that there was evidence that rape was an evolved adaptation.

According to Thornhill and Palmer, the most commonly accepted theory about rape is that it is a crime of violence, not a sexual crime. According to this theory, men learned to rape and committed the act for the purpose of exerting power and domination. But there were reasons to doubt this, the two authors said. In their view, there was a great deal of evidence that indicated that the propensity to rape had evolved either because men who raped enhanced their reproductive success by having more offspring or because it was a by-product of psychological adaptations that produced sexual desire in males.

The idea that rape is, in some sense, something that is "in our genes" was one that some critics objected to quite violently. As a result, a controversy erupted. Unfortunately, the controversy obscured the fact that Thornhill and Palmer's book wasn't just about rape. It was also an endorsement of the principles of evolutionary psychology and an attack on its opponents.

I will have occasion to comment a little more on some of the things Thornhill and Parker said a little later. But first I think it would be a good idea to comment on evolutionary psychology in general. After all, the program of this new discipline is itself controversial. If the principles of evolutionary psychology are not valid—and some scientists insist that they are not—then doubt is cast on all the ideas I have been talking about, including theories about the evolution of human sexuality, the existence of cheater-detection circuits in the brain, and evolutionary theories of rape.

Criticisms of Evolutionary Biology

Evolutionary psychology has a number of critics. Again, the most vocal of these has been Stephen Jay Gould. Gould has been criticizing evolutionary explanations of human behavior for decades, expressing his skepticism long before the discipline of evolutionary psychology even existed. He began his attacks after the publication of entomologist Edward O. Wilson's book *Sociobiology* in 1975. In this book, Wilson maintained that it should be possible to study the evolutionary roots of the behavior of social animals, whether they were ants or bees or human beings. Wilson said that if an adaptive behavior was present in all members of a species, then this behavior must have been created by natural selection. "Sociobiology," Wilson said "is defined as the systematic study of the biological basis of all social behavior." It was Wilson's aim to create a new scientific discipline.

Wilson's ideas were immediately criticized by Gould, by Richard Lewontin, and by some other scientists. They did not argue with Wilson's ideas of using the new sociobiological methods to study other animals, but they did object to the extension of these ideas to human beings. If a human behavior was adaptive (i.e., if it increased biological fitness), Gould said, then it did not necessarily follow that this behavior had a genetic origin. In humans, he argued, adaptation could come about by the nongenetic route of cultural evolution. Since cultural evolution was so much more rapid than genetic evolution, he said, its influence should prevail over any "genetic programming."

Gould's view was not unlike the one that anthropologists and sociologists have generally held. He did not deny that there were genetic factors that influenced human behavior, but he believed that these factors influenced potentialities more than they influenced specific behaviors. "What is intelligence," he asked, "if not the ability to face problems in an unprogrammed . . . way?" Most of the behavioral "traits" that Wilson and other sociobiologists were trying to explain, he said, might never have been subject to direct natural selection at all. Trying to link specific behaviors to genes, Gould concluded, was "biological determinism."

Sociobiology was not the same thing as evolutionary psychology. The latter was developed many years later. But there are links between the two disciplines. Both sociobiologists and evolutionary psychologists have attempted to explain behavior in terms of natural selection. The differences are that sociobiology is the study of the behavior of social animals in general, while evolutionary psychology deals only with human beings. Also, sociobiologists sometimes tried to explain the adaptiveness of behaviors in modern settings, while evolutionary psychologists are careful to place great emphasis on the idea that one must try to understand why typical human behaviors would have been adaptive for our hunter–gatherer ancestors.

But the two disciplines are similar, and Gould has criticized them both. That he should have done so is not very surprising. Gould consistently argues against very reductive approaches to evolutionary biology, and sociobiology and evolutionary psychology are certainly reductive. The sociobiologists spoke of specific behavioral "traits" while the entire discipline of evolutionary psychology is based on the concept of specialized mental modules. Gould, on the other hand, is a "holist" who believes that reductionism is not sufficient, that interesting new phenomena emerge at higher levels of complexity. Thus Gould has tended to see the human brain as a general-purpose computer, where the evolutionary psychologists view it as something that is made up of a distinct number of evolved components.

Although Gould has criticized the orthodox Darwinists—those he calls "Darwinian fundamentalists"—he tends to defend the older, orthodox views about the shaping of human behavior. There is nothing inconsistent about this. In both cases, he adheres to the idea that a complex system has properties that cannot be explained by analyzing the system into its fundamental components. The fundamental components are the genes in the case of evolutionary biology and mental modules in the case of evolutionary psychology. Though Gould speaks of emergent properties of complex systems only rarely, it appears that his philosophy of doing science has features in common with the philosophies of the complexity scientists.

The Debate Continues

The evolutionary psychologists have responded to Gould's criticisms on numerous different occasions. Some of the harshest criticisms of Gould were made by Thornhill and Parker in their book *A Natural History of*

Rape. The two authors took issue with the criticisms Gould had made of sociobiology and also charged that Gould had actually misrepresented contemporary evolutionary biology. Echoing Dennett, they spoke of the "unscrupulous tactics Gould has used in his attempts to discredit the evolutionary explanations that Gould apparently thinks threaten certain political ideologies." They claimed that Gould's writings had had a harmful effect, encouraging social scientists to think that they could ignore evolutionary biology as something that was not relevant to an understanding of human behavior. Gould's literary activities had "appealed to some biologists who, on ideological grounds, take the position that evolution applies to all life except human behavior and psychology." This was a "bizarre" position, they said, one that was positively "pre-Darwinian."

Other scientists have been more restrained in their criticisms. In his book *How the Mind Works* (Norton, 1997), evolutionary psychologist Steven Pinker was more moderate. Though he mentioned Gould in numerous different places, he did not engage in any direct criticisms. Instead, he was content to argue that there was a great deal of evidence to support the idea that the human mind was made up of a number of distinct modules and that these modules could be understood in evolutionary terms. However, in a 1992 article entitled "Natural Language and Natural Selection," written with psychologist Paul Bloom, Pinker engaged in a lengthy critique of Gould's views.

Although Pinker's and Bloom's topic was the evolution of language, they devoted nearly eight pages to a discussion of Gould's ideas in an effort to refute them. Pinker and Bloom began with a discussion of Gould's and Lewontin's paper on spandrels, making the point that it didn't make any difference whether a trait originally arose as a spandrel. If natural selection modified the spandrel for some adaptive purpose, the process was very much like the evolution of a trait that had been created by natural selection in the first place. The two authors then discussed Gould's and Lewontin's charges that biologists often engaged in "Panglossian" adaptationism, creating "just-so" stories that were often no more than fantasy. Natural selection, Pinker and Bloom countered, was the only explanation for the existence of adaptive traits. For example, no process other than natural selection could have produced the eye. And like other evolutionary traits, eyes exist because they conferred adaptive advantages on the organisms that possessed them. Adaptationism, they said, was not the unreasonable procedure Gould and Lewontin had made it out to be. It was an essential part of evolutionary biology.

Pinker and Bloom went on to say that furthermore Gould and Eldredge had not really shown that evolution was not gradual when they created their theory of punctuated equilibrium. In the first place, they said, the sudden appearance of "new" species in the fossil record did not necessarily imply that there had been any rapid inflationary "spurts." Many biologists believed that this could easily be explained in another way. The fossils of the "new" species would suddenly appear if it had evolved elsewhere and had then migrated into the area where the fossils were found. Next, they pointed out, some traits could only have evolved gradually. Returning to the example of the eye, they said it was inconceivable that eyes could have evolved suddenly. The only possible way in which complex eyes could have evolved was by small, incremental steps.

Only after they had discussed Gould's ideas did Pinker and Bloom speak about their main topic, the evolution of specific cognitive mechanisms underlying language. Linguistics they said, had shown that language was a complex system of many parts, a system that had been honed by natural selection for efficient communication. Explanations of the mechanisms of language were not Panglossian just-so stories, they insisted.

Human languages had characteristics that showed evidence of design, and this design could have been produced only by natural selection.

Of all Gould's criticisms, the one that has concerned evolutionary biologists the most has been the charge, made by Gould and Lewontin, that evolutionary biologists often create adaptationist just-so stories that have no basis in fact. Evolutionary psychology is based on the idea that adaptive explanations can be found for human mental traits. Thus if the ideas expressed by Gould and Lewontin in their paper on spandrels and Panglossian adaptationism were valid, then questions would be raised about the validity of the entire evolutionary psychology program. It is true that Gould and Lewontin wrote their paper in 1979, before the field of evolutionary biology was created. However the application of their ideas to questions about evolutionary psychology is obvious.

Cosmides and Tooby are among the evolutionary psychologists who have responded. In an article written for a book entitled *The Adapted Mind* (Oxford University Press, 1992), which they edited with anthropologist Jerome H. Barkow, Cosmides and Tooby stated, "The reason why Lewontin and Gould's accusation . . . that adaptationism consists of post hoc storytelling has so resoundingly failed to impress practicing evolutionary biologists is that they saw on a daily basis that adaptationism was anything but post hoc." Furthermore, Cosmides and Tooby went on, the explanation of a fact by a theory could not be post hoc if that fact was not known before the theory suggested it. Evolutionary psychology, Cosmides and Tooby claimed, had explained facts that were previously unknown; theorizing about the modules that might be present in the human mind had led to lines of research that produced new knowledge. In any case, they concluded, there was in fact nothing wrong with adaptationist thinking, and they quoted Ernst Mayr as saying, "The adaptationist question, 'What is the function of a given structure or organ?' has been for centuries the basis for every advance in physiology." Adaptationist principles, Cosmides and Tooby insisted, could provide "equally powerful guidance for research in psychology as well."

Salvos from the Opposing Camp

In 1998 Edward O. Wilson published a book titled *Consilience*, in which he discussed the prospects for unification of human knowledge, in particular the unification of the biological sciences and the humanities. "Consilience" was his term for such a unification. Wilson's book was not about sociobiology or evolutionary psychology. However, it drew on ideas developed in both those fields. Furthermore, Wilson was the founder of sociobiology. Thus it was not surprising that his book evoked responses from members of the other camp. In particular, Niles Eldredge and Stephen Jay Gould published articles about the book in the same issue of the magazine *Civilization*. The articles were intended to supplement one another.

Eldredge's contribution was ostensibly a review of Wilson's book. However, Eldredge devoted more space to making disparaging comments about evolutionary psychology and about authors such as Richard Dawkins than he did to discussing the book he was supposedly reviewing. He began the second paragraph of his review by saying,

> So we find "evolutionary psychologists" like Stephen Pinker telling us that it matters not to the end result how parents rear their children— even though anyone who had ever been a kid knows otherwise. And Richard Dawkins, of "selfish gene" fame, recently appeared in a BBC Horizon film, *Darwin's Legacy,* telling his viewers that Hitler had given eugenics a bad name.

Eldredge then linked Dawkins's selfish genes and sociobiology together by describing sociobiology as a "brilliant, if skewed, theory that described the biological world as an epiphenomenon of a mad race between genes jockeying for position in the world."

Only then did Eldredge begin to speak about Wilson's book. Wilson claimed to be trying to integrate biology with the humanities, Eldredge said, but his intent was really quite different: "the 'reduction' of the humanistic fields into the ontology of evolutionary genetics." This kind of reduction just wouldn't work, Eldredge went on. Complex systems clearly had properties of their own, properties that couldn't be explained by reductionist methods.

After making some more references to Richard Dawkins, Eldredge returned to the subject of "Wilson's raid on the humanities," and rhetorically asked, "What . . . can the evolutionary history of the human gene have to do with human culture?" He then expounded on his own views, insisting again that the behavior of large-scale systems could not be reduced to the workings of their components. Eldredge concluded by saying, in reference to Wilson's idea that systems of ethics derived, in part from our evolutionary history and our biology:

> I shudder when I hear Darwin's beautiful idea of natural selection mangled when it is applied simplistically as a moral of how we do and should behave. . . . He [Wilson] is really not so far away from the darker side— as when Richard Dawkins tells us on television that Hitler gave eugenics a bad name.

If the readers of this review came away wondering what on Earth Wilson's book was about, they probably cannot be blamed. Eldredge used the review primarily as a platform for his own ideas. In fact, he did so rather forcefully. Using some of the language of the sciences of complexity (e.g., properties of complex systems), he had argued that the reductionist program of such biologists and Dawkins, and of scientists like Wilson and the evolutionary biologists, simply would not work. To be sure, Eldredge's "review" contained some sneering references, to Dawkins in particular. But of course by this time, the debate between the two camps had become quite heated. Eldredge was not the first to have spoken in this way.

Eldredge then yielded the floor to Gould. Gould's article, which immediately followed Eldredge's "review," was titled "In Gratuitous Battle." He began by commenting on the "phony war" between the sciences and the humanities and emphasized that "the sciences and the humanities cannot be in conflict because each encompasses a separate and necessary part of human fulfillment." Then, without making any references to Wilson, Gould launched into a discussion of "the classical error of reductionism." Like Eldredge, he made references to concepts developed in the sciences of complexity such as "emergent properties" and "nonlinear interactions."

Gould stated,

> I can't think of an Earthly phenomenon more deeply intricate (for complex reasons of evolutionary mechanism and historical contingency)— and therefore more replete with nonlinear interactions and emergent features—than the human brain.

Gould then went on to his main point. Admitting that human behaviors such as cooperation might have conferred Darwinian advantages under certain circumstances and that symmetrical faces might have indeed been a sign of freedom from genetic blemishes that would hinder reproductive success, he insisted that "no such factual findings can give us the slightest clue as to the morality of morals or the esthetics of beauty." In other words, the findings of the evolutionary psychologists provided no evidence to suggest that human systems of ethics and aesthetics had a genetic foundation. Furthermore, humans were able to free themselves from genetic constraints. "We may choose to insist on cooperation even if aggression confers immediate Darwinian benefit on individuals," he said.

Finally, returning to his original topic, again making no reference to Wilson, Gould said,

> The humanities cannot be conquered, engulfed, subsumed or reduced by any logic of argument, or by any conceivable growth of scientific power. The humanities, as the most glorious emergent properties of human consciousness, stand distinct and unassailable.

Gould's article may have succeeded better as a literary essay than it did as a statement of his scientific beliefs. However, it contained arguments against the validity of the kinds of arguments employed by such scientists as Wilson and the evolutionary psychologists. By characterizing the humanities as products of the emergent properties of the human brain, Gould was implying that they were an aspect of human behavior that could *not* be reduced to mental modules or explained by reductionist methods. And his essay did complement Eldredge's contribution. Since Eldredge had already tackled the scientific issues, Gould was free to indulge himself by becoming more philosophical and making full use of his excellent literary style. Many of the readers of that issue of *Civilization* must have come away with the feeling that, whatever it was that Wilson had said, ethics and humanistic values were really not threatened by the efforts of such reductionist scientists as Richard Dawkins and Edward O. Wilson.

Gould's article in *Civilization* was not an especially clear statement of his objections to the methods of evolutionary psychology. But perhaps he did not really need to outline these objections in detail. He had already done this in his essay "Evolution: The Pleasures of Pluralism," which had appeared in *The New York Review of Books*. After replying to Dennett's charges against him, Gould had turned to a discussion of evolutionary psychology, claiming,

Humans are animals and the mind evolved; therefore all curious people must support the quest for an evolutionary psychology. But the movement that has commandeered this name adopts a fatally restrictive view of the meaning and range of evolutionary explanation. "Evolutionary psychology" has, in short, fallen into the same ultra-Darwinian trap that ensnared Daniel Dennett and his confrères—for disciples of this new art confine evolutionary accounts to the workings of natural selection and consequent adaptation for personal reproductive success.

Gould then reviewed what he considered to be the three major claims of evolutionary psychology, which he called "modularity," "universality," and "adaptation." Each of these claims, Gould said, embodied a serious weakness. Evolutionary biologists claimed that the brain could be divided into distinct modules. However, evolutionary psychologists used this concept to "atomize behavior into a priori, subjectively defined, and poorly separated items." In fact, their approach was the opposite of that of neruobiologists, who used the concept of modularity "to stress the complexity of an integrated organ."

The stress that the evolutionary psychologists placed on the universality of certain aspects of human behavior, Gould went on, was laudable. However, their new approach to universals followed the "old strategy of finding an adaptationist narrative (often in the purely speculative or story telling mode) to account for genetic differences built by natural selection."

Finally, Gould said, much of evolutionary psychology had turned into a "speculative search" for reasons why a behavior that may harm us now must once have originated for adaptive purposes. Evolutionary psychologists recognized that behavior that might have been adaptive in an ancestral environment might not be adaptive today. However, this had led to speculation that was unsupported by any evidence. For example, the human sweet tooth had supposedly arisen in an environment where fruit existed and candy didn't. A fondness for sweet things, the evolutionary psychologists said, was once adaptive, but it led to obesity today. But this was "pure guesswork in the cocktail party mode," Gould objected. No neurological evidence of a brain module for sweetness or paleontological data about ancestral feeding had been presented.

Much of evolutionary psychology, Gould charged, had turned into a search for ideas about how certain traits might have evolved in ancestral environments. However, no one really knew what these ancestral environments were. "But how can we possibly know in detail what small bands of hunter–gatherers did in Africa two million years ago?" he asked. Paleoanthropologists had discovered some bones and tools that our ancestors had left. But few inferences could be made from such evidence. Scientists did not know anything about our ancestors' concepts of kinship, about the size and structure of the groups in which they lived, the relative roles or males and females, or about "a hundred other aspects of human life that cannot be traced in fossils." The strategies used by evolutionary biologists, Gould charged, were "untestable, and therefore unscientific."

Cosmides and Tooby wrote a letter to the editor protesting Gould's characterization of evolutionary psychology. However, since Gould had made explicit reference in his essay only to Dennett and to a book on evolutionary psychology by journalist Robert Wright titled *The Moral Animal* (Cosmides' and Tooby's book *The Adapted Mind* was mentioned only in a footnote), *The New York Review of Books* printed only Dennett's and Wright's replies (they also published a reply by Pinker in a subsequent issue). Cosmides and Tooby, however, have made their unpublished letter available on their Web site, so their arguments in response to Gould are available to anyone who wishes to consult them.

Cosmides and Tooby began their long, unpublished letter by again quoting John Maynard Smith as saying that Gould's ideas were "so confused as to be hardly worth bothering with." They speculated about "the evident pain Gould is experiencing now that his actual standing within the community of professional evolutionary biologists is finally becoming more widely known." Gould's reputation as "a credible and balanced authority about evolutionary biology," they went on, "is nonexistent among those who are in a professional position to know." However, since he had misrepresented the theory to the nonprofessional public, it was necessary to make a reply.

Gould's exposition of evolutionary biology, they charged, was misleading. Furthermore, he used various kinds of rhetorical devices to misrepresent the ideas of his opponents, thus persuading his readers that they must adhere to absurd views. Gould, they went on, engaged in "minitheatricals carefully staged for the purpose of self-aggrandizement rather than for the careful and charitable pursuit of truth."

In the remainder of the letter, Cosmides and Tooby took Gould to task on specific details of his essay. Gould had said that evolutionary psychologists had investigated "the ability to detect infidelity and other forms of prevarication." This was a gross error, they claimed. What had been investigated was "an enhanced cognitive ability to reason about instances of compliance and noncompliance in situations of reciprocal cooperation." Gould had criticized evolutionary biologists for engaging in sloppy adaptationist reasoning. But, in reality, Cosmides and Tooby said, they tested not only an adaptationist hypothesis but also six different by-product hypotheses in their experiments on human reasoning. Gould's blanket condemnations of evolutionary psychology, they suggested, did not stand up when the experiments were looked at in detail.

In reply to Gould's accusation that evolutionary biologists made up just-so stories, Cosmides and Tooby repeated an argument I have cited before. Theories in evolutionary psychology had suggested previously unknown facts. Evolutionary psychologists were not engaging in post hoc and unfalsifiable storytelling. It was "predictive utility," not dogma that led evolutionary psychologists to devise the kinds of theories they did.

Cosmides and Tooby then returned to their charge that Gould had given his readers an inaccurate picture of the state of evolutionary biology. Gould had "nearly made a career" of accusing biologists of depending excessively on Panglossian explanations for the existence of evolutionary traits. But Gould knew and apparently trusted his readers not to know that "the revolution in evolutionary biology that began in the 1960s was rooted . . . in a widespread reaction against and rejection of the practice of overattributing adaptation." Gould, they charged, "customarily reverses the truth in his writing." Gould, they said, attacked fictional targets because he wanted to create a fictional image of himself as an eminently admirable individual who was "the voice of humane reason against the forces of ignorance, passion, incuriosity, and injustice."

Charges and Countercharges

The debate between Gould and his colleagues on one side and the orthodox Darwinists and the evolutionary psychologists on the other has become increasingly heated. Each camp has accused the other of misrepresenting its views, and charges and countercharges have been hurled back and forth. Gould has been described as a would-be revolutionary and has been accused of intellectual dishonesty, the orthodox Darwinists have been characterized as dogmatists, and evolutionary psychologists have been accused of sloppy thinking. Gould has been called "confused," Dawkins has been quoted out of context, and it has been implied that he endorses the aims of the discredited "science" of eugenics. Dennett has been described as "Dawkins's lapdog," and Wilson has been found to inhabit the "darker side."

The level of invective has risen to that seen in highly negative political campaigns, and the scientific issues being debated have sometimes been obscured. So perhaps it is time to cease describing the salvos that each side has periodically fired and to try to see how some of these arguments might be settled. In other words it may a good idea to try to see if any new discoveries are being made that might have a bearing on one or more of these issues and to discuss what the relevance of current research in evolutionary biology and related fields might be. I attempt to do this in the next chapter.

The
Evidence

When controversies exist, scientists look for new evidence that will resolve them. Thus any discussion of the controversies in evolutionary biology would be incomplete if it did not contain a discussion of discoveries that might have a bearing on the issues under contention. I propose to do precisely that next. I hope you won't be disappointed that the evidence sometimes seems inconclusive. Some new discoveries are relevant to some of the issues I have discussed, but others are likely to be argued about for some time because scientists do not yet know enough to enable anyone to decide one way or the other. The material I present, in other words, will be a mixed bag. Sometimes the available scientific evidence supports one side or another of a controversy. But often some points are far from being settled. I will sometimes be saying, "Yes, we now know that this idea is likely to be correct," and at other times I will be saying, "No one really knows."

Before I go into the issues currently in contention, I would like to embark on a little detour and describe some of the things that have been discovered about the processing of language in the brain. By doing this, I think I can better illuminate some of the problems that scientists currently face, especially those that relate to evolutionary psychology. Then, after I discuss evolutionary psychology, I propose to go back to some of the issues that were brought up earlier, the ones relating to spandrels, species sorting, punctuated equilibrium, and so on. I hope you will find most of these discussions interesting even though you may, at times, feel just as puzzled as many of the scientists who are trying to draw conclusions about some of these ideas.

Language Processing in the Brain

Aphasia is partial or total loss of language ability following brain damage. Its most common cause in young patients is accidental brain damage, while in older people it is usually caused by stroke. Since language is controlled by the left hemisphere of the cerebral cortex in 97 percent of all right-handed individuals and in at least 60 percent of left-handers, aphasia usually results from damage to certain areas of the left hemisphere of the cerebral cortex.

The first step toward understanding aphasia was taken in 1861 when the French physician Paul Broca dissected the brain of a deceased patient called Tan. The patient had been given that name because "tan" was the only syllable he uttered. Broca discovered a lesion in the left hemisphere of Tan's brain. When he performed dissections on the brains of an additional eight aphasic patients, he found that their brains also contained lesions in the left hemisphere.

The region of Tan's brain where Broca discovered a lesion is now known as Broca's area, and the kind of aphasia caused by damage to it is called Broca's aphasia. Few patients with this type of aphasia experience an almost-complete loss of language as Tan did, but they do tend to exhibit slow, ungrammatical speech. However, it is not possible to conclude that Broca's area is the area of the cerebral cortex that handles grammar. Severe aphasia does not occur unless some of the surrounding areas are also damaged. And some kinds of grammatical ability remain unaffected. Neurologists know that Broca's area has something to do with the production of grammatical language, but it is not possible to say precisely what its role is.

Broca's area is connected to another region of the brain, Wernicke's area, by a bundle of nerve fibers. Damage to Wernicke's area also produces aphasia, but aphasia of a different sort. Patients with Wernicke's aphasia typically speak fluently and grammatically, but their speech often does not make very much sense. Here are two examples of speech by a patient called "Mr. Gorgan" that are given by Stephen Pinker in his book *The Language Instinct* (Harper, 1994):

> Boy, I'm sweating, I'm awful nervous, you know, once in a while I get caught up, I can't mention the tarripoi, a month ago, quite a little, I've done a lot well, I impose a lot, while on the other hand, you know what I mean, I have to run around, look it over, trebbin and all that sort of stuff.
>
> Oh sure, go ahead, any old think [sic] you want. If I could I would. Oh, I'm taking the word the wrong way to say, all the barbers here whenever they stop you it's going round and round, if you know what I mean, that is tying and tying for repucer, repuceration, well, we were trying the best we could while another time it was with the beds over there the same thing . . .

Patients with Wernicke's aphasia have trouble naming familiar objects. Asked to identify a spoon, they might say "fork." They may call an elbow a knee or refer to butter as "tubber." The patients also experience difficulty understanding the speech around them. Unlike Broca's aphasia patients, they retain an understanding of grammar, but their use of language is impaired in other ways.

There are yet other kinds of aphasia. For example, if the connections between Broca's area and Wernicke's are damaged, patients find themselves unable to repeat sentences they hear. If the brain damage affects the surrounding areas of the brain, patients can repeat what they hear but show no sign of understanding the sentences they repeat.

Evidence indicates that structures in the brain perform specific tasks relating to the use of language. Thus there is some evidence that the "mental modules" of which the evolutionary psychologists speak do exist, at least where language is concerned. Furthermore, the only possible explanation for the existence of these structures is that they were created by natural selection. Although we cannot look back in evolutionary time to see how or why these structures evolved, it is certainly possible to conclude that the possession of language conferred adaptive advantages on our evolutionary ancestors. Human language is a extraordinarily complex thing. It has been estimated that the average high school graduate knows approximately 45,000 words. Choosing the correct word to use in a sentence is not a simple task. Words are made of units called morphemes, which must be combined in certain ways. Words can have suffixes and prefixes. Languages have tenses, inflections, and cases, and strings of words are combined in various kinds of phrases. And languages have unwritten rules of which we are not consciously aware. For example, as Stephen Pinker points out, in English we can say things like "Phila-fuckin-delphia," but saying "Philadel-fuckin-phia" would "get you laughed out of the pool hall."

No one really knows how language abilities are distributed throughout the brain. There may be neural circuits in Broca's and Wernicke's areas and possibly in other parts of the brain that perform various kinds of language subtasks. But if there are, no one has been able to find them. The techniques that have been developed for studying the brain are not advanced enough to find them. It is possible, for example, to look at brain activity with PET (proton emission tomography) scans. PET scans can allow scientists to determine what general areas of the brain are most active at any given time. But the blurry-looking pictures that PET scans produce give little or no information about the existence of subunits within larger sections such as Broca's and Wernicke's areas, if indeed such subunits to exist.

Studying the symptoms of patients with brain damage does not produce any evidence of subunits either. Two patients with lesions in the same general area of the brain may produce speech that is impaired in different ways, and two patients with lesions in different areas may have the same impairment. For example, about 10 percent of the people with brain damage in the vicinity of Wernicke's area have kinds of aphasia that resemble those seen in patients with Broca's area lesions. The reverse can also be the case.

Thus the evidence that neurobiologists and other scientists have found does not tell us whether the mental processes underlying language are distributed over moderately wide areas of the brain or whether they depend on specific circuits in specific locations. The evidence tells us whether the sections of the brain devoted to language function as general-purpose computers or whether they are made up of a number of distinct components. It is conceivable that most of the language processing is done in Broca's and Wernicke's areas. It is also conceivable that brain circuits for dealing with language are scattered over the cerebral cortex in a variety of different locations and that Broca's and Wernicke's areas contain processing centers that coordinate their activities. And of course it is possible to invent variations of these ideas. As I write this, it seems possible to say only that the brain does perform a large number of different tasks related to the use of language. But no one knows precisely how it does this.

Brain Research and Evolutionary Psychology

The evolutionary psychologists postulate that the human brain contains "neural circuits" that have evolved for specific adaptive reasons, for example, Cosmides' and Tooby's "cheater detector," which evolved because it is useful to us in social interactions. There is some evidence that this idea may be correct, at least as it applies to language. But knowledge of language processing in the brain is too incomplete to draw very many conclusions. Scientists do not know how many language centers there are in the brain or how they perform specific tasks. Most of their knowledge comes from studies of patients with brain damage, and the evidence is often hard to interpret. Recall, for example, that patients with lesions in Broca's area sometimes exhibit symptoms that look very much like those of Wernicke's aphasia.

Neurobiologists have gained considerable knowledge about the brain, but their findings have generally not proved to be very relevant to evolutionary psychology. They know, for example, that a brain structure called the hippocampus is essential for the production of memory, but they are uncertain of the precise way in which memories are formed. A large gray mass of nerve cells called the amygdala is associated with feelings of fear, but other parts of the brain also probably contribute to evoking fear. Scientific knowledge of the brain is woefully incomplete. Scientists do not know how the brain acquires and stores information, how it produces feelings of pleasure and pain, or how it creates consciousness. The functioning of the human brain is a profound mystery, one that scientists have only begun to understand.

We have to conclude, then, that the idea of "neural circuits" in the brain that have evolved to perform specific tasks has not been confirmed. This does not imply that evolutionary psychology is bad science. Cosmides, Tooby, and other evolutionary psychologists believe that they have discovered indirect evidence that allows them to conclude that the brain does indeed contain such circuits. In their view, the cognitive functions that have been discovered are so specific that brain modules of some kind must exist.

The critics of evolutionary psychology do not agree. Gould, for example, continues to insist that the brain does not operate in this manner. In his view, most of our mental abilities are spandrels that were not created by natural selection. Gould has given reading and writing as examples. Since written language is a relatively recent invention, it could not be an ability that evolved during ancestral times on the African savannas. The ability to read and write must be a by-product of the way the brain is constructed. Indeed, it would be easy to construct quite a large list of human intellectual abilities that could not have been shaped by natural selection. Such a list might include such things as the ability to learn higher mathematics, to understand complicated games like chess, to play a violin, and perhaps even to form such linguistic constructions as "Phila-fuckin-delphia."

But what about genetic evidence? As many as 30,000 different genes may be involved in the construction of the human brain. Is it possible to relate specific genes to certain behavioral traits? Sometimes reports in the news media give the impression that genes have been discovered that are related to such abnormal conditions as Alzheimer's disease, schizophrenia, and alcoholism. It is sometimes easy to get the impression that the discovery of genes "for" other behaviors and mental conditions is imminent. It has even been reported that scientists have discovered a single gene that somehow "controls grammar." But these accounts are misleading. Most genetic traits are the product not of single genes but of networks of genes. It is usually impossible to say that a gene has been discovered "for" some trait.

In his book *The Language Instinct,* Stephen Pinker gives an excellent example. Discussing the so-called "grammar gene," he points out that all that is known is that a certain defective gene (actually the gene was not identified; is was presumed to exist because it caused a syndrome than ran in families) that is found in some individuals is thought to disrupt grammar. In no sense does it control grammatical speech. If the disruptive gene is present, it creates a condition that is somewhat analogous to a car with defective spark plugs. Without operating spark plugs, the car will not run. But it doesn't follow that the spark plugs somehow "run" the car. It wouldn't operate if the fuel line was cut or if the wheels were removed, either.

There is much about the brain that is not well understood. Scientists know even less about the role played by genes in creating the structures found in the brain. Thus it seems to be necessary to conclude that the assumptions made by the evolutionary psychologists are not based on any significant quantity of neurological or genetic evidence. But we can't conclude that evolutionary psychology is based on mistaken assumptions. The neurological and genetic evidence does not provide any good arguments against evolutionary psychology, either.

Evolutionary psychologists are scientists who are attempting to extend the frontiers of human knowledge. They certainly cannot be criticized for trying to do this. And there is every reason to think that they will make some very interesting discoveries. Some of these discoveries may not be what they anticipate. They may be able to confirm their theories, but it is possible that they will discover evidence that indicates that the human mind is largely a collection of spandrels, as Gould claims. It is not likely that the controversies evoked by the claims evolutionary psychologists have made will be resolved soon.

But this isn't necessarily so bad. Controversies aren't unhealthy things. They often serve to delineate issues. If scientists lack the evidence they need to decide whether specialized neural circuits such as those postulated by Cosmides and Tooby do indeed exist, they at least know what they should be looking for. If anything is certain, it is that in the years ahead neurobiologists will make important discoveries about the functioning of the brain. There are more than enough reasons to attempt to devise new experimental techniques and to apply them to new lines of research. And when new discoveries are made, it is likely that new light will be thrown not only on the problems of evolutionary psychology but on questions raised in many other areas of scientific endeavor, as well.

Levels of Complexity

Evolutionary psychologists have made some very specific claims about mental modules. It is at least possible to imagine how some of these claims might be verified or disproved, even though scientific knowledge about the brain is too incomplete to do so at present. In the case of other controversies, matters are not always so clear. For example, Eldredge and Gould seem to think that the theories of the orthodox Darwinists are too reductionist, that they place too much emphasis on natural selection alone. A reductionist approach, they say, neglects phenomena that may be taking place at a higher level of complexity. They sometimes refer to the work that has been done in the sciences of complexity and speak of emergent properties of complex systems.

You should not should not automatically conclude that their "holistic" approach is necessarily superior to the reductionist methods of such scientists as John Maynard Smith and Richard Dawkins. In many areas a reductionist approach is better, and in some circumstances it is the only reasonable way to proceed. For example, it is not likely that physicists would have succeeded in understanding the behavior of elementary particles if they had tried to approach the problem "holistically." They succeeded because they were reductionists, who analyzed atoms into their component parts and discovered that such heavy particles such as protons and neutrons were made up of even smaller particles called quarks. If scientists had continued to look at the "holistic" properties, as the alchemists did, the science of chemistry would not exist. That science was created when scientists began to analyze chemical compounds into their components and to develop rules that described the ways in which chemical elements combined with one another.

Eldredge and Gould maintain that natural selection is not the only important factor in evolution. In particular, they appeal to the idea of species sorting. But it is difficult to determine whether species sorting really takes place or how important a factor it is in evolution. One of the problems has to do with time scales. Natural selection over periods of decades has been observed in such animals as the peppered moth and the house sparrow. They did not evolve into new species, but they underwent changes that could be observed. Species sorting, on the other hand, is a process that would take place over periods of many thousands of years. Since this is much longer than the lifetimes of individual scientists, species sorting would be difficult or impossible to observe directly. Most likely some kinds of species sorting take place. However, at the very least, it is difficult to say how important it is in the general scheme of things.

Complexity scientists have created models of evolution. Since evolution inside a computer proceeds so much more rapidly than it does in the real world, their work might conceivably throw some light on long-term trends in evolution and help to resolve some of the controversies I discuss in this book. Unfortunately, this research is only beginning, and few of the features of real evolutionary systems have been reproduced. For example, punctuated equilibrium has been seen in Tom Ray's Tierra simulations, but its appearance depends on the kind of computer language that is being used. Digital evolution is slow and gradual in one simulation and proceeds by sudden leaps in another. To be sure, Ray's Tierra is not the only computer simulation in which evolution has been observed. However, it is the most advanced in the sense that more time and work have been expended on it than any of the others.

There are also questions about how much light computer simulations can shed on real biological phenomena. Complexity scientists have produced useful simulations of such things as traffic patterns, financial markets, the collective behavior of ants and honeybees and ecological networks. However, no one has as yet attempted to produce simulation of something as basic as a biological cell. Admittedly, Tom Ray's multicellular digital organisms are made up of "cells." However they are hundreds of thousands or millions of times simpler than complex eukaryotic biological cells, and they are not intended to simulate all the processes that take place within the cells of biological organisms. Ray's multicellular organisms had six genes. By comparison, the simplest bacteria have approximately a thousand. Ray's multicellular "ancestor" was constructed of some 640 computer instructions. The human genome is estimated to contain well over 100 megabytes of information.

Of course Ray wasn't attempting to simulate the behavior of cells. He wanted to see how modestly complex digital organisms would evolve. However, a comparison between the electronic cells he created and real cells does illustrate the gap that currently exists between models created by complexity scientists and organisms in the real world.

Stuart Kauffman has created computer models of the origin of life and of gene networks. He has investigated the question of the evolvability of such networks and has reached conclusions that might be relevant to the evolution of biological life. He has made discoveries about the degree of interconnectedness that a gene network presumably should have if natural selection is to have any chance of modifying organisms in adaptive ways. Kauffman's work may be important. But no one knows whether or not the models that he has created are an accurate reflection of reality. Scientists know relatively little about the ways in which gene networks create biological traits. To date, the complete genomes of only a few organisms have been mapped, and no one has begun to try to understand the details of how the numerous genes in an organism interact with one another.

It may be that the sciences of complexity will contribute a great deal to scientific understanding of the problems of biology in the future. Biological organisms and patterns in evolution are obviously complex, and they certainly exhibit emergent properties of some kind. However, scientists do not as yet have a good understanding of what these emergent properties are. Eldredge and Gould have emphasized that, in their opinion, these properties must exist. According to Gould, this makes a "fundamentalist" (i.e., purely reductionist) approach to evolution invalid. But this question seems to be far from settled.

Spandrels

In their 1979 paper, Gould and Lewontin charged that evolutionary biologists engaged in Panglossian speculation and relied excessively on just-so stories to explain the structures found in living organisms. They attempted to reductively break organisms down into "traits" and then explain how each trait could have been created by natural selection. To bolster their argument, the two authors introduced the concept of spandrels, traits that arose as a by-product of natural selection, which served no evolutionary purpose.

The orthodox biologists denied that they engaged in sloppy practices. The creation of adaptive explanations, if carried out carefully, was good science. No one could deny, for example, that eyes were for seeing or that possessing vision caused organisms to be better adapted to their environments. In numerous other cases, it was obvious what observed traits were for and why they had evolved.

Gould and Lewontin were not arguing with other evolutionary biologists about scientific facts. Their criticism was a methodological one, and the publication of their paper certainly made biologists more aware of the need to be rigorous when they talked about the purpose served by some structure found in an organism. So the paper probably had a salutary effect. Anyone who points out that a certain amount of rigor is needed when one frames scientific arguments is certainly performing some small service to the scientific community.

Gould and Lewontin made no major criticisms of evolutionary theory when they wrote about the overuse of adaptationist stories. Their paper would most likely have been quickly forgotten except for one thing: they introduced the idea of spandrels. The concept of spandrels quickly took on a life of its own, causing Gould's and Lewontin's paper to be cited again and again in arguments about evolutionary theory.

Spandrels certainly exist. In the paper he wrote with Lewontin, Gould was able to point out some features seen in various species of marine animals that did not seem to serve any useful functions. But it is not apparent that this fact is of any great significance. As the orthodox evolutionary biologists pointed out, natural selection could and did modify spandrels for adaptive purposes. Or, as Stephen Pinker put it, "We evolved from organisms without eyes, feet, and other complex organs. The organs must have originated in precursors that were spandrels for some ancestral organisms."

When Pinker said this, he was repeating the often-heard objection that *of course* spandrels existed and *of course* natural selection operated on them. He was saying, in effect, "Lewontin and Gould say there are spandrels, but so what?" Gould and Lewontin, on the other hand, admit that spandrels are modified by natural selection. It would seem that there shouldn't be any continuing argument.

But there is an ongoing argument. It has to do with Gould's conception of the human brain. Gould has said:

> Natural selection made the human brain big, but most of our mental properties and potentials may be spandrels—that is, nonadaptive side consequences of building a device with such structural complexity.

And Pinker has replied:

> Evolutionary psychologists are not ignorant of this hypothesis. They have considered it and have found it to be unhelpful.

This, I think, is the real issue. Gould believes that the brain is a complex thing that has many capabilities, most of which are there as a result of its great complexity and size. Evolutionary psychologists like Pinker believe that the brain can be broken down into components that evolved for specific purposes.

This argument is not likely to be settled soon. It's resolution will depend on finding the same kinds of evidence that would validate or invalidate the claims of such evolutionary psychologists as Cosmides and Tooby. If they are eventually able to show that the brain contains a large number of mental modules that were created by natural selection, then they will also have shown that Gould was wrong, that most of the properties of the brain are not spandrels. On the other hand, if at some point neuro-biologists find evidence that shows that the brain works as an integrated whole as a kind of general-purpose computer, that it cannot be broken down into groups of neural circuits that evolved to perform specific tasks, it will probably be necessary to conclude that Gould was right. The question of the existence or nonexistence of large numbers of spandrels in the brain cannot be separated from the problem of determining whether the fundamental outlook of the evolutionary psychologists is correct.

I can't help but think that the question of spandrels is not a very important one as it pertains to other areas of evolutionary biology. In spite of their sometimes heated rhetoric, the two sides do not really seem to be very far apart. Both sides agree that spandrels exist, and both sides agree that it is foolish to try to find an adaptive explanation for every trait that is observed in an organism. Gould and Lewontin would probably argue that spandrels play a more important role in evolution than the orthodox Darwinists think they do. But the difference in their outlooks is really not so great. Gould is less of a reductionist than his opponents; he argues that it is sometimes wrong to break an organism down into too many different "traits." But, again, the difference is only a matter of degree. Gould has given adaptive explanations of certain biological traits himself. He and his colleagues do not oppose the use of adaptive arguments in general; they only caution that such arguments are sometimes overused.

Punctuated Equilibrium

When it comes to the theory of punctuated equilibrium, we find our-selves on somewhat firmer ground. Questions of the validity of reduction-ism versus a more "pluralistic" approach to evolution tend to seem at least partly philosophical in nature. When we try to decide how seriously we should take the claims of evolutionary psychology, it is hard to resist the temptation to hesitate, scratch our heads wondering whether the evolutionary psychologists or their opponents should be believed, and then decide that it will probably be necessary to wait and see what evidence will be discovered in the future. Gould's and Eldredge's theory of punctuated equilibrium, on the other hand, is based on solid paleontological evidence. That evidence may be subject to various different interpretations. But at least it is possible to point to observed facts when arguing about the significance of the theory.

Furthermore, new evidence is being discovered that may have a bearing on the theory. Several different groups of scientists have discovered evidence that indicates that evolution is not always as gradual as biologists, Eldredge and Gould included, have generally assumed it to be. They have found that observable evolutionary change can take place quite rapidly, in some cases more rapidly than had been thought possible. Even Eldredge and Gould thought it would probably take tens of thousands of years for new species to evolve. But recent published studies, though they do not contain observations of any new species, provide evidence that speciation might not require that much time.

When I say that, I am engaging in speculation, of course. The scientists who have observed rapid evolution have, in a typically cautious scientific manner, contented themselves with stating the facts. It would be going too far to say that they have discovered anything that proves that speciation is rapid. But I can't help but think that they have opened up the possibility. And the fact that they have at least gives hope that biologists will soon discover evidence that gives them a better understanding of the process of speciation.

But before I describe this new evidence and speculate further about its possible implications, it might be best to briefly recapitulate a bit and summarize some of the main points made by Eldredge and Gould and by their critics.

According to the theory of punctuated equilibrium, species typically change very little over long periods of time, which may be as long as several million years. These periods of stasis, the theory says, are punctuated by short periods of rapid evolution, which are perhaps 10,000 to 50,000 years in length. Furthermore, most evolution is associated with speciation, the creation of new species. In other words, evolution takes place when new species are evolving; the rest of the time, natural selection acts to keep species stable.

According to Eldredge's and Gould's hypothesis of species sorting, natural selection acting on individual organisms cannot fully account for all evolutionary change. Other processes, they say, take place at higher levels. For example, some species will survive and will produce new daughter species in turn, while others go extinct.

Some scientists, such as Stephen Pinker, have clung to the traditional idea that, when a new species suddenly appears in the fossil record at some geological location, this should be interpreted to mean that this species migrated into the area. However, as Eldredge has pointed out, species cannot always be evolving elsewhere. Consequently, most biologists accept the idea that punctuated equilibrium is valid to some extent. Although some emphasize that gradual evolution may take place between punctuational bursts, they generally agree that the short bursts of evolution are real.

But orthodox Darwinians such as Richard Dawkins and Daniel Dennett deny that the theory of punctuated equilibrium is anything but a minor correction to evolutionary theory. Evolution that takes place over a period of tens of thousands of years, they say, still has to be considered to be gradual. Furthermore, they add, evolutionary biologists have always been aware of the fact that evolution did not always proceed at the same rate. Gould and Eldredge, they concluded, didn't introduce any very revolutionary idea; their theory can easily be accommodated within orthodox evolutionary theory.

Furthermore, they were skeptical of the idea that species sorting could be of any great significance. It was natural selection acting on individual organisms, Dawkins emphasized once again, that was responsible for evolutionary change. Species sorting was nothing more than a kind of group selection. At best it could be of only minor importance.

But as they went on arguing, other scientists were going out in the field to do research.

New Evidence

In 1999, two different groups of scientists discovered new evidence that indicated that evolution could take place more rapidly than had previously been thought. At the same time, their results suggested that natural selection was indeed as powerful as the orthodox Darwinists had always insisted, that it was invariably the driving force behind evolutionary change.

Some evolutionary biologists have believed that not all evolutionary change could be attributed to natural selection, that something called "genetic drift" might also play a role. Genetic drift is a term that was coined to describe random changes in the genetic makeup of populations of organisms. There is no guarantee, the proponents of this idea said, that the organisms with the best genetic endowment would always be the ones that survived. Chance also plays a role. No matter how well adapted an organism is to its environment, ill luck could cause it to die of injury or disease before it reproduced. This would lead to random gene changes, they suggested, which would accumulate over long periods of time, perhaps giving rise to new species.

However, the two studies published in the journal *Science* in early 2000 suggested that perhaps this is not the case, that natural selection is indeed as important as Darwin and contemporary orthodox Darwinists had thought. The first of these studies was carried out by biologist Raymond Huey of the University of Washington, Seattle, and his colleagues. They studied a European fruit fly that had been introduced into California some twenty years previously. In 1997 they collected flies in some eleven different locations ranging from just north of Santa Barbara, California, to north of Vancouver, British Columbia. The following year Huey and Spanish colleagues trapped flies at locations from southern Spain to the middle of Denmark that lay at roughly the same latitudes.

Huey and his coworkers then bred the flies from each location for a half-dozen generations, giving them all the same food and providing them with the same living conditions. They did this to eliminate any differences in the flies that might have been caused by differences in local environments. They wanted to determine how flies collected at lower, warmer latitudes differed from those found at higher, colder ones. Although they are at roughly the same latitude, Santa Barbara and southern Spain do not provide anything like identical conditions. And of course British Columbia is different in many ways from Denmark.

Finally, they measured the wing lengths of the flies. They saw a difference of about 4 percent between the flies collected at different locations on a given continent. The flies that had been bred from those collected at more northerly locations had longer wings. Furthermore, the differences were the same in both European and North American flies. No wing-size differences had been observed in North American flies collected a decade previously. The North American wing-size difference had apparently evolved in just ten years. Furthermore, the wing size differences were the same as those that had evolved in European flies.

The flies with longer wings also had slightly larger bodies, and it is possible to guess that the increase in body size in more northerly locations was an adaptation to cooler climate. However, the cause of the differences was not of any particular concern to Huey and his coworkers. The results suggested something far more important: that evolutionary change could take place more rapidly than anyone had imagined. They also indicated that it was natural selection—and not genetic drift—that had caused the differences. In fact, natural selection had found the same adaptive solution to the problem of changing climate in North America that it had in Europe.

One should not assume that the genetic makeups of European and North American flies changed in the same ways. They probably didn't. In the European flies, the part of the wing closest to the body was lengthened, while the North American flies showed an increase in size in the part of the wing farthest from the body. Natural selection had apparently found two different genetic solutions to the problem of lengthening wings in Europe and in North America. There had been parallel evolution, but the paths taken seemed to be different.

A second study, carried out by biologist Dolph Schluter of the University of British Columbia and his colleagues, also dealt with parallel evolution. Schluter and his colleagues collected stickleback fish from three lakes in British Columbia. These fish, which were originally of marine origin, had been trapped in these lakes approximately 10,000 years ago, and had been isolated from one another ever since.

The researchers found that the same two species of stickleback had evolved in all three lakes. Each lake contained a species of larger, bottom-dwelling fish and also smaller, more streamlined stickleback that lived and fed in the open water. In 10,000 years, natural selection had caused each population to split into two species, and the two species had similar characteristics in all three lakes.

In each lake, the two species were reproductively isolated; they did not interbreed with one another. So three UBC scientists, Laura Nagel, Janette Boughman, and Howard Rundle, decided to determine the mating preferences of female fish from different lakes. They found that the females chose males that looked like themselves, whether they came from the females' own lakes or not. For example, bottom dwellers would mate with bottom dwellers from a different lake and would not mate with fish from their own lake that were unlike themselves. In other words, natural selection had caused the fish to evolve along parallel courses to the extent that the same mating preferences had evolved in the fish in all three lakes. The fact that bottom dwellers and open-water swimmers in a single lake had been separated from one another for a shorter time than similar fish from different lakes seemed to make no difference.

The results of these studies were significant for several reasons. They also demonstrated the overriding importance of natural selection. Under similar conditions, the same two species had evolved in each lake. Furthermore, they had evolved along parallel lines. Genetic drift had apparently not played any important role. If it had, fish from different lakes would not be so similar to one another that they would mate. Furthermore, the studies showed that, in at least some cases, new species arose for reasons other than geographical isolation. It was not the separation of the lakes from one another than had given rise to new species but rather the fact that each lake contained two ecological niches to which the sticklebacks could adapt themselves.

Although only three species were studied in these two research programs, they seem to provide some confirmation of Eldredge's and Gould's idea that evolutionary change can take place rapidly and that change is associated with speciation. In all three lakes, new species arose is very short periods of time, periods of time shorter than the 10,000 to 50,000 years sometimes cited in discussions of the theory of punctuated equilibrium. On the other hand, the two research studies seem to give some support to a point that the orthodox Darwinists have always emphasized: that natural selection is the sole significant cause of evolutionary change. You should note, however, that I say "some support." The two studies both showed that there had been parallel evolution, and they both suggested that natural selection had been the major cause and that other factors hadn't played much of a role, if any. In particular, genetic drift seemed to have been ruled out. But two studies are not enough to allow one to make generalizations about the entire biological world. It was still possible that factors other than natural selection could have a role to play under different conditions.

And of course the studies were not conducted to settle any controversies. Two teams of scientists conceived research programs in the hope that these programs would help to provide a better understanding of natural selection and of evolution. In the end, they gave some support to a major tenet of Maynard Smith, Dawkins, and other orthodox Darwinists, while also providing evidence that could be construed as suggesting that some of the ideas of Gould and Eldredge had been correct. But they didn't provide any final answers.

The Ecologists Return

Until fairly recently, evolutionary theory was dominated by such mathematical geneticists as John Maynard Smith. Mathematical genetics deals with such matters as the change of gene frequencies under the influence of natural selection. You probably won't be surprised if I tell you that this emphasis on genetics has tended to foster a very reductionist view of evolution. Maynard Smith, Dawkins, and their colleagues attribute all evolutionary change to natural selection. Dawkins has gone even further: he has claimed that natural selection not only provides an explanation for all evolutionary change, but that it should also be capable of giving us a better understanding of the interactions between species in ecological systems.

However, recent research indicates that the opposite may be the case, that ecological studies may give us an understanding of some of the ways in which new species are created. According to University of California at Berkeley biologist James Patton, "ecology has been out of favor [among evolutionary biologists] for over two decades." But now that is changing. Scientists who are trying to understand the details of how one species splits into two are beginning to pay attention to ecologies in great detail. They are observing such things as the number or predators that lizards are likely to encounter in different forests and the angles at which bottom-dwelling fish feed. Their findings have caused them to revive some ideas about speciation that have long been out of favor.

As I have noted previously, orthodox Darwinists tend to adhere to the theory that speciation generally takes place when two segments of a population of organisms are isolated from one another, for example, by a mountain range. If the two environments are different, then natural selection will drive them apart. And even if the two environments were identical, genetic drift would cause them to evolve in two different random directions. But if the two subpopulations were not isolated, their thinking goes, then they would interbreed, and no new species would arise.

However, in recent years, an alternative theory has begun to gain ground. It isn't a new theory, but it is one that has only recently begun to be tested extensively. According to this theory, the barriers that separate nascent species can be ecological rather than geographic. Research that has been done in the field has shown that this can indeed be the case. We have already encountered an example of this. In each of the Canadian lakes that Schluter and his colleagues studied, one species of sticklefish had evolved into two. In each of the lakes, the two species had different body shapes, and females of one species would not breed with the other.

The theory doesn't say that ecological isolation is the only cause of the creation of new species. It is generally admitted that geographical separation can lead to speciation, just as the orthodox Darwinists claim. The advocates of the ecological model simply say that new species can arise in more than one way. This claim has been documented by studies in numerous different locations. Among the species studied are finches on the Galápagos islands, Trinidad guppies, and rainforest birds known as little greenbuls that are found in Cameroon. Some of these studies show that evolutionary change can be quite rapid when populations are separated by ecological rather than geographic barriers. For example the little greenbuls were observed to evolve longer wings in just twenty years when they migrated to more open forest. And the cichlid fish that inhabit Lake Victoria in Africa have evolved into hundreds of species in just 12,000 years.

As I write this, Dolph Schluter and his colleagues have been conducting a new study of stickleback fish, comparing marine and freshwater fish in British Columbia to counterparts in Japan that live in similar environments. The results he obtained were similar to those of his study of sticklebacks in the three British Columbia lakes. He found that the freshwater sticklebacks on both sides of the Pacific looked almost identical to one another, and that they both were quite different from ancestral marine forms.

In an experiment in which Shluter collaborated with Japanese and American scientists, the mating preferences of the fish were again studied. It was found that both Japanese and Canadian females preferred to mate with male fish that looked like themselves. Japanese marine females would mate with their Canadian marine counterparts but not with stickle-backs that came from a freshwater environment in Japan. The Canadian freshwater females showed similar preferences; they would mate with Japanese freshwater fish, but not with Canadian marine sticklebacks. Furthermore, these matings produced viable hybrids.

The significance of this study was this: it showed that extreme geographical isolation had not caused the fish to diverge from one another to the extent that they became reproductively isolated. But ecological separation did erect mating barriers. Geographical separation had done little or nothing to create new species. This was one case where the orthodox theory about the way in which new species were created was clearly wrong.

More Challenges to Orthodoxy

According to orthodox Darwinian theory, natural selection produces small variations. Some organisms have genetic makeups that make them slightly more likely than other members of their species to survive and reproduce, ensuring that, over time, their genes will spread throughout a population. Furthermore, according to orthodox thinking, evolution takes place through the slow accumulation of small, favorable mutations.

Until recently, this idea was widely accepted, and its acceptance caused evolutionary biologists to believe that evolutionary change was something that happened over long periods of geological time. Eldredge and Gould challenged the idea with their theory of punctuated equilibrium, but their opponents replied that evolutionary change that took place over periods of tens of thousands of years still had to be considered to be gradual. The theory of punctuated equilibrium, they said, had done little to modify long-accepted ideas.

In some of the studies I have cited, scientists have found that evolutionary change can be surprisingly rapid. For example, the study of North American fruit flies showed that measurable differences in wing size could arise in as short a period of time as a decade. However, the genetic differences between flies with different wing lengths were not studied. Consequently little could be said about the genetic causes of wing-size variation either in North America or in Europe. However, other scientists are studying the genetics of species differences, and they have recently made some surprising discoveries.

When graduate student Corbin Jones of the University of Rochester studied two related species of fruit fly in the Seychelles archipelago of the Indian Ocean, he and his adviser, evolutionary geneticist H. Allen Orr, expected that he would find a big genetic difference between them. The evolution of a new species, after all, was supposed to be caused by a large number of genetic changes, each of which had a small effect.

The two flies that Jones studied had the Latin names *Drosophila simulans* and *Drosophila sechellia*. *D. sechellia* lived and laid its eggs on the Morinda fruit, a foul-smelling fruit that was poisonous to most insects. *D. simulans,* on the other hand, avoided the Morinda and fed on a variety of other fruits. It looked like a classic case of two species separated by an ecological barrier.

But when Jones carried out a genetic mapping of the genes of the two species of flies in 1998, he discovered that there were only a few differences between them. A small handful of genes gave *D. Sechellia* resistance to the Morinda fruit's poison, and an equally small number seemed to account for the fly's attraction to the Morinda's scent. The difference between the two species could be accounted for by a few genes that caused big behavioral changes.

This was a surprise because big, beneficial mutations were thought to be so rare that they played no significant role in evolution. But it was apparently not a fluke. At the about the same time Jones was carrying out his research, other biologists were obtaining similar results. For example, in 1996, evolutionary biologist Doug Schemske and geneticist Tom Bradshaw of the University of Washington, Seattle, performed a genetic mapping of bee-pollinated and hummingbird-pollinated monkey flowers in California's Yosemite National Park. They discovered that the two species of flowers differ from one another in only a small number of genes. But these few genes have large effects; they cause differences in flower color, petal shape, and quantities of nectar produced. These differences were responsible for luring the two different kinds of pollinators and keeping the species reproductively isolated.

Evolutionary biologists aren't quite sure what to make of these results. They clearly contradict generally accepted theory. Big mutations were supposed to have been very rare and mostly disadvantageous. But here were some data that showed that the opposite could be the case. The data did seem to show that adaptive change could take place very quickly; a change that involves only a few genes can come about more rapidly than one that depends on small changes in a lot of genes. But there was no existing theoretical framework in which these results could be explained. Commenting on this situation, Orr said, "We're in a funny situation—we're about to have a new wave of data crash down on us and no theory to hang it on."

Some of the scientists who have performed studies of this type have begun to speculate that a new model of evolutionary adaptation may be needed. According to Schemske, it may be that large mutations sometimes happen first and are followed by a series of small mutations that fine-tune a species to its ecological niche. Orr has followed up on this idea by creating a mathematical model that attempts to show how a big-adaptations-first pattern would work. This model, which has been published in the journal *Evolution,* is intended to provide a theoretical framework for the kinds of results that Jones, Schemske, and Bradshaw have obtained. Orr's theory attempts to describe the role that big, rapid mutations play in evolution. Naturally, it contradicts orthodox Darwinian theory, which begins with the assumption that only small mutations have any significant role to play.

Ongoing Research

Evolutionary biology is not a closed subject. More research studies are being carried out than ever before, and scientists learn something new about evolution every day. The studies that I have cited represent only a small number of those that are currently being done. For example, ecological studies of speciation have been carried out not only with sticklebacks, fruit flies, and monkey flowers but also with rainforest lizards, Australian snails and bats, and Amazon River rodents. And new studies are being planned constantly. Scientists are testing ideas about speciation in every imaginable way.

The pace of research has increased in other biological fields as well. Organic chemists, ecologists, geneticists, developmental biologists, and other scientists are seeking to gain greater insight into the patterns of evolution. Scientists are studying the evolution of bacteria in the laboratory. Advances are being made in the field of paleontology, and scientists are looking at the evolution of evolutionary diversity in rapidly reproducing bacteria in the laboratory. For example, Richard Lenski of Michigan State University has been breeding and experimenting on bacterial cultures since 1988. Since his experiments began, the bacteria have gone through more than 24,000 generations, more than enough time for a considerable amount of evolution to take place (by comparison, 24,000 human generations would amount to something like 500,000 years).

Meanwhile, scientists are learning more about early life—organisms that existed over 2 billion years ago—and have developed some fascinating new theories. For example, physicist Norman Sleep of Stanford University and planetary physicists of NASA's Ames Research Center have theorized that life may have originated on Earth when an impact with an asteroid knocked off a chunk of the planet Mars and sent it hurtling toward Earth. If life did evolve on Mars, it most likely evolved earlier than it could have on Earth; the surface of Mars had already cooled considerably when that of Earth was still molten.

Admittedly, the possible Martian origin of life does not have very much to do with the controversies I have been describing. And neither does some of the other research that is being carried out. But these examples do illustrate the fact that scientists are not only literally probing every corner of Earth in order to gain a better understanding of evolution; they sometimes look beyond Earth as well.

Since so much work is being done, it is not surprising that some of it should appear to throw new light on Eldredge's and Gould's theory of punctuated equilibrium. In particular, if speciation can take place when two populations of a species begin to inhabit different ecological niches in the same locale, then some of the sudden appearances of new species that paleontologists find in the fossil record represent something real and are not the results of organisms migrating from elsewhere. If evolution can sometimes proceed as rapidly as some of the studies I have cited seem to indicate, then perhaps new species can appear even more suddenly than Eldredge and Gould and their critics have estimated.

As far as I know, none of the research currently being done throws much light on Eldredge's and Gould's idea of species sorting—or of the significance of spandrels, for that matter. However, some long-accepted ideas about evolution are being questioned. For example, there seems to be evidence that evolution does not always proceed by the slow accumulation of small mutations. Contrary to what had previously been believed, sudden large mutations seem sometimes to play a role. And if this particular phenomenon turns out to be common, then the mathematical geneticists who have dominated evolutionary theory are going to have to rethink some of their ideas. No one can say what the picture of the evolution of life on Earth will be ten years from now. But it is likely to be different in some respects from the one scientists have today.

Controversy
and Discovery

Within a decade after *On the Origin of Species* was published in 1859, the great majority of British biologists had accepted the idea that evolution had indeed taken place. And yet even Darwin's staunchest supporters failed to agree with Darwin on various different points. Although they believed that there was convincing evidence for evolution, they did not all accept the idea that natural selection was the cause of evolution or that evolutionary change was as gradual as Darwin claimed.

The controversies subsided during the 1930s when such mathematical geneticists as J. B. S. Haldane, Sewall Wright, and Ronald Fisher showed that Mendel's genetics could be reconciled with Darwin's theory. They used mathematical arguments to demonstrate that if natural selection acted on small variations that had accumulated over a period of time, then significant evolutionary change would take place. The work was influential because it was formulated in mathematical language and because it showed that mutations alone were not sufficient to explain evolutionary trends that took place over long periods of time. However, the arguments were initially very theoretical in nature; they had little empirical support. Nor did the theory seem able to take speciation into account. To be sure, some experiments were performed, mostly with fruit flies. But the laboratory animals tended to be highly inbred, and laboratory environments were not much like the flies' natural habitats. Mutations were often induced by subjecting the flies to X rays, which was hardly the kind of thing they would encounter in the natural world.

Then, in 1937, a book by the Russian-American naturalist and geneticist Theodosius Dobzhansky, titled *Genetics and the Origin of Species,* remedied some of these defects. Dobzhansky had studied fruit flies in nature, rather than in the laboratory, and was thus able to provide real experimental evidence in support of the new theory. Dobzhansky also dealt with such topics as speciation, and he was thus able to give a fairly comprehensive account of evolution.

A great deal of research has been performed since 1937. Experimental evidence has accumulated, new theoretical ideas have been proposed, and new disciplines, such as molecular biology, have been created. However, there is one thing that has not changed. Orthodox Darwinian theory is still based on mathematical genetics. By its very nature, this theory is reductionist, and its continued acceptance has bolstered the idea that natural selection, acting on the individual, is the sole cause of evolutionary change.

Since the 1930s, mathematical genetics has been so dominant that paleontologists and other evolutionary scientists have contributed little to evolutionary theory. To be sure, a great number of important discoveries have been made. During the last sixty or seventy years, discoveries of new fossils have contributed enormously to our understanding of the history of life. As 1 explained in earlier chapters, molecular biologists have deciphered the genetic code, and techniques have been developed for comparing the DNA of different species. Genetic evidence has given scientists a better understanding of lines of evolutionary descent. However, the theory of why evolutionary change takes place was a creation of the geneticists, and the geneticists have retained their hold on theory to this day.

In 1984 John Maynard Smith published an article in *Nature* in which he welcomed paleontologists back to the "High Table" of evolutionary discourse. Too often, Maynard Smith said, the "attitude of population geneticists to any paleontologist rash enough to offer a contribution to evolutionary theory has been to tell him to go away and find another fossil, and not to bother the grownups." The welcome was short-lived, however. When scientists such as Stephen Jay Gould, Richard Lewontin and Niles Eldredge engaged in criticisms of accepted evolutionary theory, the geneticists sometimes proved to be more concerned about defending their turf than they were in engaging in meaningful dialogue. As the debates described in this book grew hotter, Eldredge began to describe the geneticists and certain of their allies, like Richard Dawkins, as "ultra-Darwinists." A few years later, Gould coined his own term, "Darwinian fundamentalists." Maynard Smith meanwhile labeled Gould's ideas as "confused," and Daniel Dennett launched an all-out attack on Gould.

As the argument intensified, the number of participants increased and the charges and countercharges flew furiously. The debate widened to include evolutionary psychology and the issues under debate were obscured as those on each side accused those on the other of misconstruing and misrepresenting what they had said.

The Role of Scientific Controversies

Many lay people conceive of science as a body of knowledge that has become more or less firmly established. To some extent this is true. For example it is not likely that anything short of divine revelation is ever going to convince biologists that evolution did not happen. The likelihood that scientists will ever discover that life has not been evolving on Earth for at least 3.5 billion years isn't much greater than the chance that we will eventually find that Earth does not revolve around the sun. But science is something else as well. It is a way of gaining new knowledge about the universe in which we live. Scientists don't spend their professional lives preaching dogma. On the contrary, they apply themselves to problems that are as yet unsolved. The "golden ages" of science are not times when there seems to be little more to be learned; they are times in which mysteries and unanswered questions abound.

It is inevitable that, during such times, scientific controversies arise. Sometimes there is not enough evidence to settle a question, and sometimes the evidence that has been discovered so far is open to more than one interpretation. Under such circumstances, controversy can be a healthy thing. As issues are debated, they are thrown into clearer focus, and scientists gain a better understanding of the kinds of evidence that must be found is the debates are finally to be laid to rest.

I don't mean to imply that scientific controversies are always settled. Sometimes they are not. Sometimes they fade away. It may be that, as the complexities of a problem are uncovered, it turns out that there was some truth on both sides of the issue. Alternatively, ideas that are hotly debated turn out to be irrelevant. For example, during the latter part of the nineteenth and the beginning of the twentieth century, physicists debated vociferously about the properties of the ether, the hypothetical substance that filled all space and that carried light waves from the sun to Earth and other planets. But the arguments quickly ceased after Einstein's theory of relativity showed that the idea of the ether was superfluous.

In other cases, new discoveries are made that suddenly make one of the two opposing positions seem untenable. This is what happened in mid-nineteenth century England, for example, when the idea of evolution itself was being debated. The geologists, paleontologists, zoologists, botanists, and other scientists rapidly amassed evidence that demonstrated that Darwin must have been right. To be sure, there were some doubters well into the twentieth century. But the majority of them either grew old and died, or finally became convinced when the synthesis of Darwin's theory with genetics was achieved.

None of the controversies I have described in this book have yet been resolved. However, it some cases, it is already clear how they can be. For example, as neurobiologists and cognitive psychologists amass more knowledge about the structure and workings of the human brain, the hypothesis that the brain contains a great number of evolved, special-purpose neural circuits is likely to come to seem either more or less plausible. This hypothesis can, in principle, be proved or disproved, although there is not enough evidence to do so yet.

If this controversy is settled, the result will have implications for Gould's and Lewontin's theory of spandrels. Although these two scientists did not originally invent the concept with the intention of applying it to the functioning of the human brain, the question of the existence of spandrels has turned out to be a hotly debated issue. The problem of whether the brain is composed mostly of spandrels or of specialized neural circuits is closely related to numerous questions about the manner in which the human intellect functions.

The other issues involving spandrels may not be terribly important ones. Recall that Gould and Lewontin originally introduced the idea to emphasize the fact that every trait of an organism need not serve an adaptive function. Some of them exist because they are by-products of other traits that were molded by natural selection. Gould, Lewontin, and their critics all agree that, once a spandrel is created, natural selection may modify it for some useful purpose. In any case, Gould and Lewontin did not criticize any commonly accepted theoretical ideas in their original paper. The main point they made was about methodology. They asserted that stories about the supposed adaptive origin of this or that trait were sometimes ad hoc. Evolutionary scientists, they said, sometimes invented explanations of why certain traits had evolved that sounded plausible but were not supported by any empirical evidence. This, Gould and Lewontin said, was not good science. Their paper elicited a great deal of argument as to the extent to which orthodox Darwinians really did spin Panglossian stories. But even if they didn't to the extent that Gould and Lewontin implied, the warning was still a useful one. It never hurts to insist on strict scientific procedures.

The Problem of Reductionism

Eldredge and Gould have consistently attacked the reductionism of the mathematical geneticists and of scientists like Dawkins who would reduce everything in evolution to natural selection. Eldredge and Gould have a more "pluralistic" view of evolution. While admitting the importance of natural selection, they insist that there are also processes that take place at higher levels. In doing so, they have raised important questions about the validity of the reductionist program.

In particular, as you probably recall, Eldredge and Gould have introduced the idea of species sorting, or competition between species. They consider it to be a higher-level process that contributes to evolutionary change, although not to the degree that natural selection does. There has been a great deal of debate about the question of whether or not this idea is valid. But there is little evidence that would support either their claims or those of their detractors.

I think the debate about reductionism in evolutionary biology is only beginning. It is an issue that has been raised in many different areas of debate. For example, Gould has charged that evolutionary biology has become too reductionist. He has criticized the tendency of some biologists to break organisms down into a large number of distinct "traits." He has expressed skepticism about the claims of the evolutionary psychologists that the human mind can be broken down into "mental modules" and that human behavior is in large part a collection of behavioral traits that have been molded by natural selection. He believes that, on the contrary, biological organisms and the human mind function as integrated wholes.

Gould is not alone in insisting on a more "holistic" approach. Eldredge has expressed similar ideas, and so have the researchers who work in the sciences of complexity, a discipline based on the idea that reduction does not always work, that it is often necessary to look at the properties of entire systems instead.

Stuart Kauffman in particular has been seeking to understand the ways in which the self-organizing properties of complex systems might play a role in evolution. Although he has produced an imposing body of work, it is not yet possible to test his theories in the laboratory. However, if his ideas turn out to be correct, it will be necessary to conclude that certain kinds of biological order arise spontaneously. Kauffman has also advanced some hypotheses about the ways in which natural selection and self-organization might work together to produce evolutionary change. His theoretical studies of gene networks suggest that some types of networks have better "evolvability" than others and that natural selection will therefore cause them to proliferate. There are questions about the extent to which Kauffman's simplified models reflect processes that take place in the natural world. However, the work he has done so far is certainly an interesting beginning.

Meanwhile, there is promise that the research of Thomas Ray and other scientists in evolving systems inside computers. will eventually lead to new insights about the nature of the evolutionary process. In particular, they may gain new knowledge about the evolution of complex organisms. But, again, they are only beginning. There are still problems to be solved about the nature of evolvability. In particular they do not yet know what characteristics of evolving systems depend on the computer environment and which arise from the evolutionary process itself. However, their experiments have already reproduced phenomena, such as the evolution of parasites and gene duplication, that are seen in the biological world.

Many biologists have remained skeptical about the relevance of the sciences of complexity to their discipline. But this is only natural. They have not yet become used to the idea that self-organization of complex systems may play an important role in the biological world. Most of them have continued to adhere to a very reductionist outlook, and the ways of thinking that seem so natural to complexity scientists are somewhat alien to them.

At this point, it is only possible to make guesses about the contributions to evolutionary thought that may be made by the sciences of complexity in the future. The discipline is still young, and many of the conclusions that have been drawn by scientists working with computer models cannot yet be subjected to empirical tests. However, it will be surprising if the work in the field does not eventually change biological thinking in some ways. After all, organisms are very complex systems, and some kinds of self-organizing properties must exist. For that matter, individual cells are extraordinarily complex, and reductionist methods don't come anywhere near giving us an understanding of everything that goes on within them.

It appears that there is only one conclusion that can be drawn with any degree of confidence: no one really knows what lies ahead, but whatever it turns out to be, it is likely to surprise us.

New Evidence

Recently, some surprising new discoveries about evolution have been made, and new evidence has been uncovered that may make it necessary to modify some long-held theories. As I said in Chapter 8, it has been found that traditional ideas about the causes of speciation are wrong, at least in some cases, that evolution can proceed at a more rapid rate than anyone had suspected, and that natural selection is indeed all-powerful.

The studies that led to these discoveries were not carried out to settle any of the controversies I have been discussing. They were performed simply because scientists wanted to gain a better understanding of the ways in which evolutionary change takes place. If some of them threw some light on the debates between the two contending parties, it was purely accidental. Furthermore, only a small number of species were studied, and it is not possible to extrapolate from these results far-reaching conclusions about evolutionary patterns in general. Nevertheless, these studies do seem to be significant ones that may lead to the formulation of new ideas, which can then be tested by further experiments.

The studies of fruit flies in North America and in Europe that I described in the previous chapter were surprising because they showed that observable evolutionary change could take place in as little as ten years. Furthermore, the North American flies showed the same kind of variation in wing length with latitude that was seen in European flies (which had adapted themselves to their environment long ago). Apparently, natural selection had been faced with the same problem on both continents and had found the same solution in each case. This indicated that genetic drift, which had often been cited as a factor in evolutionary change, had not been significant. The evidence indicated that only natural selection had played a significant role.

A similar kind of parallel evolution was seen in the stickleback fish that inhabited Canadian lakes that had been isolated from each other around 10,000 years ago. The same two kinds of fish evolved in each lake. Each contained streamlined sticklebacks that swam in open water and a species of bottom dwellers that were larger and had a different shape. This showed that the long-held idea that speciation was caused by geographical separation was not always correct. One species of stickleback had apparently split into two because two populations had been ecologically separated, living in different kinds of environments within the same lake.

Furthermore, natural selection was again shown to have been all-powerful. In all the lakes, the same kinds of fish evolved. Fish from different lakes were so similar that free-swimming fish from one would mate with free-swimming fish from another, while bottom-dwellers did the same. Natural selection under similar conditions had again led to parallel evolution and similarities between fish that had been geographically isolated from one another for 10,000 years were so great that they recognized one another as potential mates.

These studies seemed to show that, at least in some cases, Eldredge's and Gould's idea that evolutionary change was associated with speciation were correct. Furthermore, if ecological isolation rather than geographical separation could lead to the rise of new species, then one of the arguments made by opponents of the theory of punctuated equilibrium did not hold water. When a new species suddenly appeared in the fossil record, one need not conclude that it had migrated from somewhere else. On the other hand, again in these particular cases, there seemed to have been no evolutionary "pluralism." It appeared that natural selection had been the only important factor.

It would be a mistake to make wide-ranging generalizations that are based on the study of only a few species in particular kinds of circumstances. However, evidence concerning the importance of ecological separation in evolution seems to be accumulating.

Meanwhile, other research has led to results that may call another long-held idea into question. The studies of fruit flies on islands in the Indian ocean and of monkey flowers in California have shown that the belief that evolution always takes place through the slow accumulation of small mutations (i.e., mutations with small effects) may not always be correct. In these two cases, large mutations involving just a few genes had caused dramatic changes in the organisms that were being studied. Again, studies of just two species are not enough to permit far-reaching conclusions. However, if similar phenomena are observed by other researchers, the implications could be important indeed.

A Science in Ferment

Evolutionary biology is a science in ferment. It is a field in which new discoveries are being made at an ever-increasing rate, a field in which controversies abound. Scientists such as Lewontin, Eldredge, and Gould have been questioning certain tenets of the prevailing orthodoxy. Gould, especially, has acted as a gadfly and has aroused the ire of the advocates of Darwinian orthodoxy.

In recent decades, entire new disciplines, such as the sciences of complexity and evolutionary psychology, have been developed. It is probably too soon to tell how many of the ideas advanced by scientists who work in these fields will turn out to be valid. But it will be very surprising if the researchers who work in these fields do not contribute significantly to scientific in the years ahead. The complexity scientists have developed new ways of looking at complex systems, and the hypotheses advanced by evolutionary psychologists have suggested promising new lines of research.

Both fields have their critics. But this is likely only to motivate the scientists who work in these disciplines to be more rigorous. Being subjected to criticism is not always pleasant. But it does cause one to be careful about assumptions and to attempt to establish the correctness of conclusions beyond all doubt. And if some of the hypotheses advanced by the complexity scientists and the evolutionary psychologists turn out to be wrong, that will be no great tragedy. Science is full of discredited hypotheses. Discovering that certain ideas are incorrect has often had the effect of causing scientists to follow other, more promising roads.

Meanwhile, advances continue to be made in already established disciplines. By the time you read what I am writing, the human genome will either be completely mapped, or the project will be nearing completion. But when the task is completed, the search for knowledge will not come to an end. Once the genome of an organism is known, scientists can begin to study the proteins that the genes in that organism produce. The knowledge gained from such protein studies—which are only beginning—are likely to lead to significant new advances in molecular biology and in evolutionary biology as well. The interaction of proteins within a cell is also something that could be modeled on computers using the methods of the sciences of complexity. So increased knowledge of cellular proteins could lead to advances in that field as well.

Not all research is conducted in the laboratory. New field studies that have been recently performed have indicated that, at least in some cases, some of the tenets of orthodox Darwinian theory, especially those about the causes of speciation, are wrong. At the same time, they have given added support to the idea that natural selection is of overriding importance. It is not yet possible to tell where the lines of research I have described are likely to lead. However, it will not be surprising if it becomes possible to settle some of the controversies I have described. But if it is, evolutionary biology is not likely to become a less contentious field. As the old controversies disappear, new ones are likely to appear in their place.

Scientific controversy, after all, is a healthy thing. It can exist only when certain accepted ideas are being questioned or when there is debate over new hypotheses or newly discovered experimental data. I have said all this before, but perhaps it is worth emphasizing again that the existence of controversy is invariably a sign that a scientific field is very much alive. After all, when a body of scientific knowledge becomes so well established that little or nothing is open to doubt, then there is no longer anything to argue about.

A Selected, Annotated Bibliography

For the most part, I have cited only books that are accessible to the general reader. A few exceptions are noted.

Barkow, Jerome H.; Cosmides, Leda, and Tooby, John. 1992. *The Adapted Mind.* New York: Oxford University Press. A collection of essays on various topics in evolutionary psychology. This book is not intended for a popular audience. Nevertheless, the essays are not difficult to read, even though they are a little technical at times. In one of the contributions, Cosmides and Tooby define the goals of evolutionary psychology. David Buss and Steven Pinker are among the other contributors.

Brockman, John. 1995. *The Third Culture.* New York: Simon & Schuster. Literary agent John Brockman taped interviews with a number of leading scientists, many of whom had done work in evolutionary biology or in the sciences of complexity. He then edited the tapes to make each interview read like an essay. Also included are comments by these scientists on one another's work. Contributors include Stephen Jay Gould, Richard Dawkins, Niles Eldredge, Daniel Dennett, and Stuart Kauffman.

Buss, David M. 1994. *The Evolution of Desire.* New York: Basic Books. An evolutionary psychologist looks at human mating preferences.

Darwin, Charles. 1859. *On the Origin of Species by Means of Natural Selection, or the Preservation of Favoured Races in the Struggle for Life.* London: John Murray. The original edition. Naturally there have been numerous editions since 1859. Darwin's book is also available on line (see the list of selected Web sites that follows this bibliography).

Dawkins, Richard. 1976. *The Selfish Gene.* Oxford: Oxford University Press. Dawkins expounds the idea that evolution is a result of the fact that genes "want" to leave as many copies of themselves in succeeding generations as possible. Of course this is really only a metaphor. What Dawkins is really doing is expounding the orthodox Darwinian idea that everything can be reduced to natural selection acting on the individual.

Dawkins, Richard. 1986. *The Blind Watchmaker.* Harlow, U.K.: Longman. Dawkins discusses the role of chance in evolution and reiterates his ideas about "selfish genes."

Dawkins, Richard. 1996. *Climbing Mount Improbable.* New York: Norton. Dawkins discusses bout the evolution of seemingly improbable biological structures and behavior.

Dennett, Daniel C. 1995. *Darwin's Dangerous Idea.* New York: Simon & Schuster. An exposition of the "ultra-Darwinist" point of view, followed by speculations on philosophical topics. Dennett argues that those who will not accept "Darwin's dangerous idea" that natural selection is the basis of all evolutionary change are afraid of it because it seems to threaten humanistic values. Naturally there is a lot of commentary on the work of Dennett's *bête noir,* Stephen Jay Gould.

Depew, David J., and Weber, Bruce H. 1995. *Darwinism Evolving.* Cambridge: Cambridge University Press. This one gets somewhat technical. However, it is a comprehensive account of the evolution of evolutionary theory from Darwin's time to the present.

Eldredge, Niles. 1999. *The Pattern of Evolution.* New York: W. H. Freeman and Company. The latest book in which Eldredge explains his and Gould's ideas.

Eldredge, Niles. 1995. *Reinventing Darwin.* New York: John Wiley. The best popular account of Eldredge's and Gould's theory of punctuated equilibrium and the idea of species sorting. Eldredge explains his and Gould's belief that natural selection is not the only factor in evolution, that there are also phenomena at higher levels of complexity.

Farrington, Benjamin. 1966. *What Darwin Really Said.* New York: Schocken. A short, very readable account.

Gell-Mann, Murray. 1994. *The Quark and the Jaguar.* New York: W. H. Freeman and Company. Nobel laureate Gell-Mann discusses the two fields that have interested him the most: high-energy particle physics and the sciences of complexity. Gell-Mann writes the deepest discussion of the meaning of the term "complexity" I have run across. Among other things, he relates mathematical definitions of the term to complexity in the real world and finds that none of them are really quite adequate. This isn't very surprising. Hundreds of definitions of and measures of complexity have been proposed. However, Gell-Mann's discussions give the reader real insights into the difficulty of the problem.

Goodwin, Brian. 1994. *How the Leopard Changed Its Spots.* New York: Scribner. Goodwin expounds his ideas about biological forms that are not the result of natural selection. The subtitle, *The Evolution of Complexity,* is a reasonably good description of this book.

Gould, Stephen Jay. 1998. *Leonardo's Mountain of Clams and the Diet of Worms.* New York: Harmony Books. Gould writes a monthly column for *Natural History,* and the essays that make up his columns have been collected in numerous books that have been published over the years. This book and the following one are the two most recent in the series.

Gould, Stephen Jay. 2000. *The Lying Stones of Marrakech.* New York: Harmony Books.

Kauffman, Stuart. 1995. *At Home in the Universe*. Oxford: Oxford University Press. A semipopular treatment of the ideas expressed in *The Origins of Order* (following).

Kauffman, Stuart. 1993. *The Origins of Order*. Oxford: Oxford University Press. This has been called "a monster of a book." Even scientists sometimes have trouble following all Kauffman says, if they lack the proper mathematical background. Nevertheless, it is the most complete exposition of Kauffman's theories.

Levy, Steven. 1992. *Artificial Life*. New York: Random House. Levy discusses the development of the field of artificial life (artificial digital organisms). This is another book that is very easy to read. It provides an excellent introduction to the earlier work in the field.

Lewin, Roger. 1992. *Complexity*. New York: Macmillan. Lewin discusses the sciences of complexity and the people who did the most to develop the field. There are a lot of anecdotes about the scientists' personal lives.

Lewin, Roger. 1999. *A Revolution in Evolution*. New York: W. H. Freeman and Company. Lewin describes the development of techniques for analyzing DNA and explains how new techniques in molecular genetics have provided new insights into the evolution of life.

Maynard Smith, John. 1958. *The Theory of Evolution*. Cambridge: Cambridge University Press. An account of the orthodox theory.

Mayr, Ernst. 1991. *One Long Argument*. Cambridge, MA: Harvard University Press. The subtitle—*Charles Darwin and the Genesis of Modern Evolutionary Thought*—describes the book perfectly.

Morris, Richard. 1999. *Artificial Worlds*. New York: Plenum. A popular account of work in the sciences of complexity that relates to questions concerning the origin and evolution of life.

Pinker, Steven. 1997. *How the Mind Works*. New York: Norton. An excellent popular account of evolutionary psychology.

Pinker, Steven. 1994. *The Language Instinct*. New York: Morrow. Pinker discusses the innate components of human language.

Ridley, Matt. 1997. *The Origins of Virtue*. New York: Viking Penguin. This is a well-written popular book on evolutionary psychology. The author discusses theories about the evolution of cooperative behavior among human beings.

Ruse, Michael. 1999. *The Darwinian Revolution*, 2nd ed. Chicago: University of Chicago Press. Ruse discusses the conversion of the British scientific community to belief in evolution during the years 1830–1875.

Thornhill, Randy and Palmer, Craig T. 2000. *A Natural History of Rape*. Cambridge, MA: The MIT Press. The controversial book about the possible genetic underpinnings of rape.

Wesson, Robert. 1991. *Beyond Natural Selection*. Cambridge, MA: The MIT Press. In this extensively researched book, Wesson makes a case for the contention that natural selection cannot be all there is to evolution.

Williams, George C. 1966. *Adaptation and Natural Selection*. Princeton, NJ: Princeton University Press. A classic book. Richard Dawkins popularized Williams's ideas in *The Selfish Gene*.

Wilson, Edward O. 1998. *Consilience*. New York: Knopf. Wilson discusses the fundamental unity of human knowledge. The book was controversial because some scientists objected to Wilson's idea that the information in our genes could account for literature, art, morality and religion.

Wilson, Edward O. 1978. *On Human Nature*. Cambridge, MA: Harvard University Press. The debates over sociobiology took place over twenty years ago. Nevertheless, the questions that were raised, such as that of how much of human behavior is influenced by our genetic endowment, are still of interest. This book was largely responsible for provoking the debates.

Wright, Robert. 1994. *The Moral Animal*. New York: Random House. A comprehensive popular treatment of evolutionary psychology. This was the book that Stephen Jay Gould singled out for attack in one of his essays in *The New York Review of Books*. Although Wright is a journalist, not a scientist, he has since involved himself in the controversies centering around some of the things that Gould has said.

World Wide Web Resources

A lot of Web sites contain information about evolution and about the ideas discussed in this book. Some of the material to which I refer can be found only on line. For example, the letter by Leda Cosmides and John Tooby in which they respond to Stephen Jay Gould's criticisms of evolutionary psychology can be found only at their site. Other items are much easier to find on the Web than elsewhere. For example, I found it much simpler to go to the Web site of *The New York Review of Books* to find Stephen Jay Gould's essays on "Darwinian fundamentalism" and on his own, more pluralistic views than to hunt them down in back issues of the publication at the library.

The Web also contains a lot of information useful to anyone who wants to explore further some of the topics I discuss. Thus I do not hesitate to include a few sites that deal with topics that are only briefly mentioned in this book. I include a few sites with downloadable "artificial life" software. These software packages allow you to run simulations of animal behavior or of evolution on your computer. They're mostly pretty simple. But that's the whole idea. The science of complexity deal with complex systems whose components interact with one another in simple ways.

Boids. http://www.red.com/cwr/boids.html
A site devoted to Craig Reynolds's Boids, which are digital flocking animals. Although the boids interact with one another by following only three simple rules, their behavior eerily resembles that of real birds, so much so that

Reynolds's program has elicited the interest of ornithologists. This site provides an excellent example of the fact that complex systems can exhibit self-organizing behavior even when the number of interacting entities is small.

Boston Review. http://18.171.0.185/BostonReview/evolution.html
Gould and Eldredge are not the only scientists who have criticized the orthodox Darwinists. Some of the debate about the issues I have presented in this book can be found in *Boston Review*, published by the Massachusetts Institute of Technology. Both Dawkins and Dennett have been criticized in articles published in it. This isn't a biological journal, so the material it prints is quite readable.

Toby Bradshaw. http://poplar2.cfr.washington.edu/toby/
Bradshaw and geneticist Doug Schemske of the University of Washington at Seattle conducted a study in which they found large differences between two species of monkey flowers, even though the flowers differed from one another in only a few sets of genes.

Center for Evolutionary Psychology. http://www.psych.ucsb.edu/research/cep/
Leda Cosmides and John Tooby are Directors of the Center for Evolutionary Biology at the University of California at Santa Barbara. At this site you can find information about their current and recent research and also their reply to Stephen Jay Gould's criticisms of evolutionary psychology.

Complexity On-Line. http://life.csu.edu.au/complex/
A scientific information network about complex systems.

Daniel Dennett. http://ase.tufts.edu/cogstud/pubpage.htm/
Many of the papers and other writings of the philosopher Daniel Dennett can be accessed at a Web site maintained by the Tufts University Center for Cognitive Studies, of which Dennett is director. Most of this material is *not* about evolution or about the controversies in which Dennett has taken part. However, Dennett's interests are wide-ranging, and much of this material has great intrinsic interest. And yes, the site does contain a small number of articles on evolution and on the science of complexity.

Evolution and Behavior. http://ccp.uchicago.edu/~jyin/
A site maintained by the University of Chicago psychologist Jie Yin.

Evolution Homepage.
http://bioinfo.med.utoronto.ca/~lamoran/Evolution_home.shtml/
A huge collection of links. If there is anything you want to know about evolution, you can find it by going to this site.

Evolution Update. http://www.geocities.com/CapeCanaveral/Lab/7111/
Information about the latest news, feature articles, and books relating to evolutionary biology.

Evolutionary Psychology for the Common Person. http://www.evoyage.com/
There is a lot of material on evolutionary psychology here. This site also contains information about books on the subject.

Evolutionary Psychology Research Lab.
http://homepage.psy.utexas.edu/homepage/group/BussLAB/index.htm
David Buss is director of the Evolutionary Psychology Research Laboratory at the University of Texas

Stephen Jay Gould. Gould does not use a computer and he does not have a Web page. But his articles can be found at various other sites on the Web. For example, see The World of Richard Dawkins.

Helix. http://www.necrobones.com/alife/
Helix is a Tierra-like system for Windows. It can be downloaded from this site. Also available: BugFest, a predator–prey simulation that uses digital simulated insects.

Homo Deceptus. http://slate.msn.com/Earthling/96-11-27/Earthling.asp/
"Homo deceptus" is a reference to Stephen Jay Gould. Robert Wright, a journalist who has written a book on evolutionary psychology titled *The Moral Animal*, engages in a diatribe against Gould here. Another essay by Wright about Gould can be found at http://www.nonzero.org/newyorker.htm.

Thomas Henry Huxley Archive. http://www.vbook.org/free/THHuxley/
T. H. Huxley earned the name "Darwin's Bulldog" for his vehement defense of Darwin's evolutionary theory. Many of the essays that are available here are still worth reading. Huxley, incidentally, was the individual who coined the term "agnostic" to describe his religious outlook.

Human Behavior and Evolution Society. http://157.242.64.83/index.htm
There are lots of links here to sites with material on evolutionary psychology.

Stuart Kauffman. http://www.santafe.edu/sfi/People/kauffman/
Kauffman's home page. There really isn't a great deal of material available here. But if you're willing to wade through something that becomes a bit technical at times, you might want to look at Kauffman's "Investigations," his speculations about ideas in the science of complexity that "are not yet science."

Richard Lenski. http://www.msu.edu/~lenski/
Lenski has bred over 24,000 generations of *E. coli* bacteria in an ongoing experiment that has lasted more than a decade. By comparison, 24,000 human generations would be about 500,000 years. Thus he has been able to observe some significant biological change during the course of the experiment.

John Maynard Smith. http://www.biols.susx.ac.uk/faculty/biology/maynard.htm
Maynard Smith's home page.

McLean vs. Arkansas Board of Education.
http://cns-web.bu.edu/pub/dorman/mva.html
The decision by U.S. District Judge William R. Overton. Overton held that "creation science" was not science but religion and that it could therefore not be taught in Arkansas schools. This isn't the case reviewed by the U.S. Supreme Court, which declared laws specifying that creationism be taught alongside evolutionary biology were unconstitutional. The court reviewed a Louisiana law in that case. However, Overton's decision states the legal and scientific issues especially clearly. A number of prominent scientists, including Stephen Jay Gould, testified in this case.

Nature Science Update. http://helix.nature.com/nsu/
A news site maintained by the British Journal *Nature*. The articles posted here are nontechnical and are meant for a broad audience. News items about new findings about evolution appear here frequently.

The New York Review of Books. http://www.nybooks.com/nyrev/index.html
Essays and reviews by and reviews of books by Daniel Dennett, Niles Eldredge, Stephen Jay Gould, Richard Lewontin, Steven Pinker, and John Maynard Smith can be found at this site. The search engine allows you to find them easily. The reviews are often followed by lengthy exchanges prompted by letters from authors whose books were criticized in reviews.

Orr Lab.
http://www.rochester.edu:80/College/BIO/labs/ORRLAB/ORRHOME.HTML
The home page of H. Allen Orr, an evolutionary geneticist at the University of Rochester. It was Orr's student Corbin Jones who did the study of fruit flies in an archipelago in the Indian Ocean and discovered that small genetic differences between two fly species caused large behavioral differences. And if you think you know something about fruit flies yourself, you might want to click on Orr's "Test Your Fruit Fly IQ." There is also a link that takes you to a book review Orr wrote of Daniel Dennett's book *Darwin's Dangerous Idea*. In this review, titled "Dennett's Strange Idea," Orr (who sides with Gould) criticizes Dennett at length. For Dennett's reply to this review, go to
http://18.171.0.185/BostonReview/br21.5/dennett.html

Steven Pinker. http://www-bcs.mit.edu/~steve/
Steven Pinker's home page.

Santa Fe Institute http://www.santafe.edu/
The Santa Fe Institute in New Mexico is a research center devoted to the study of complex systems. If you're interested in finding out about some of the work currently being done in the science of complexity, this is the place to look. Artificial life software, information about current research, links to the Web pages of indi-

vidual scientists are among the items you will find. Note: Many of the papers here are available for download only in PostScript format. If you don't have PostScript, you can view and print them with Ghostview and Ghostscript, respectively. Both are free and can be found in a variety of sites. Just do a search.

Dolph Schluter. http://www.zoology.ubc.ca/~schluter/Brochure.html
Schluter is the University of British Columbia scientist who has studied stickleback fish in Canadian lakes.

Science and Culture. http://www.shef.ac.uk/~psysc/rmy/indsac.html
A site maintained by the journal *Science and Culture*. Among the items that can be found here is an article about sociobiology and evolutionary psychology titled "Sociobiology Sanitized." To find it, click on "Latest Articles and Papers."

The Talk Origins Archive.
http://www.talkorigins.org/origins/faqs-evolution.html
This site has articles on almost every conceivable topic in evolution. Many of the articles analyze the flaws in creationist theories. However, they can generally be read for their intrinsic interest if you're not especially concerned about creationist attacks on science. Topics covered include "Introduction to Evolutionary Biology," "What Is Evolution?" "Evolution Is a Fact and a Theory," "The Modern Synthesis of Genetics," and many others. There is a section on Eldredge's and Gould's theory of punctuated equilibrium and a section on human evolution. The complete text of Darwin's *On the Origin of Species* can also be found here.

Tierra. http://www.hip.atr.co.jp/~ray/tierra/tierra.html
Thomas Ray's Tierra. The original Tierra program is available for download here. It will run on Windows, Mac, Unix, and Linux computers. Many of Ray's papers are reports on the network Tierra project.

World of Richard Dawkins, The. www.spacelab.net/~catalj
Dawkins is not associated with this site. There is an enormous amount of interesting material on Dawkins and his books here. It is also an excellent place to find links to articles by Stephen Jay Gould: there is an entire section devoted to Gould, called "The Gould Files."

Zooland. http://alife.santafe.edu/~joke/zooland/
A collection of artificial life resources maintained by the Santa Fe Institute. And why does the word "joke" appear in the address? The creators of this site say, "Don't take anything too seriously; it's only science."

Adaptation, without selection, 86
Adaptation and Natural Selection
(Williams), 76–77
The Adapted Mind (Cosmides, Tooby
& Barkow), 184–185, 190
Adaptionism, 84–85, 183–185,
206–207
Adenine, in protein synthesis, 63–64
Algae, blue-green, 31
Algorithm, natural selection as,
89–92
Altruistic behavior, 171–175
Amino acids, 63–64. *See also*
Proteins
in autocatalysis, 139–140
sources of, 68–69
Amygdala, 200
Anthropological relativism, 166, 167
Antibiotic resistance, bacterial, 39
Apes
brain size in, 35
evolutionary split from humans,
71–72
evolution of, 34
genetic similarity of to humans,
34, 43, 71
Aphasia, 196–199

Broca's, 196–198, 199
Wernicke's, 197–198, 199
Artificial life, in computers, 143–147
Asteroids collisions, 25, 34, 67–68,
80–81, 223
Australopithecines, 34, 35
Australopithecus ramidus, 72
Autocatalytic systems, 137–141

Bacteria, 26
antibiotic resistance in, 39
evolutionary studies of, 223
fossilized, 25, 31
primordial, 25–26
properties of, 26
structure of, 26–27
Barkow, Jerome H., 184
Behavior, genetic basis of. *See*
Evolutionary psychology;
Sociobiology
Big-adaptations-first model, 222
Biological determinism, 180
Birds, microevolution in, 37–38
The Blind Watchmaker (Dawkins),
111
Bloom, Paul, 182–184
Blue-green algae, 31

Boughman, Janette, 215
Bradshaw, Tom, 221, 222
Brain
  current understanding of, 200–203
  Gould's view of, 208–209
  language processing in, 196–200
  neural circuits in, 168–171,
    174–175, 180–181, 200–203.
    See also Neural circuitry
  spandrels in, 201, 208–209
Brain function, natural selection and,
  170–171
Breeding
  animal, 42–43, 114
  plant, 42–43
Broca, Paul, 196–197
Broca's aphasia, 196–198, 199
Broca's area, 196–197, 199
Brocks, Jochen J., 27
Brodie, Benjamin, 48
Buffon, Comte, 16–17
Buick, Roger, 27
Buss, David, 162–165, 168

Cambrian Explosion, 30–31
Candide (Voltaire), 84–85
Carbon dating, 21
Catalysts, 137–141
Cells
  bacterial, 26
  eukaryotic, 26–27
  evolution of, 26–27
  plant, 27
Chambers, Robert, 48
Cheater detection, 174–178, 200
Chemicals, organic
  autocatalysis among, 137–141
  sources of, 68–69

Chimpanzees
  brain size in, 35
  evolutionary split from humans,
    71–72
  evolution of, 34
  genetic similarity of to humans,
    34, 43, 71–72
Chloroplasts, 27, 28
Chromosomes, mapping of, 67
Civilization, 185, 188
Coming of Age in Samoa (Mead),
  161
Common descent, 52, 67
Complexity sciences, 7–9, 125–155,
  204–206, 232, 235–236
  computer models in, 127–128,
    205–206, 232
  limitations of, 154
  punctuated equilibrium and,
    141–142
Complex systems
  catalysis in, 137–141
  computer models of, 127–128,
    127–137
  emergent properties of, 127,
    130–131, 203–206
  genomes as, 129, 132–137
Computer models, 8–9
  in complexity sciences, 127–128,
    205–206, 232
  of evolution, 141–153, 205
  of gene networks, 128–137,
    152–153, 206
  limitations of, 205–206
  of origin of life, 206
Computers
  artificial life in, 143–147
  virtual, 144

Consilience (Wilson), 185
Contingency, 119–122
Cooperative behavior, 171–175
Correns, Carl Erich, 60
Correspondence, 43–44
Cosmides, Leda, 2, 167–170,
  175–177, 184–185, 190–192,
  200, 201, 203, 208
Crick, Francis, 63
Cultural relativism, 166, 167
Cyanobacteria, 31
Cynodonts, 33
Cytosine, in protein synthesis, 63–64

Darwin, Charles, 45–46, 47, 50, 51,
  52, 54–55, 58, 59, 73, 92, 99, 225
Darwinian fundamentalism, 1–2, 5n,
  82–83, 141, 142, 202–204, 227,
  230–233
  vs. evolutionary pluralism, 1–2,
  5–6, 78–87, 92–98, 118–119,
  185–193, 203–204, 230–233
  mathematical genetics and,
  216–217
Darwinian theory
  contemporary attacks on, 47–52
  early supporters of, 50–52
  gradualism in, 51, 53, 60–63,
  99–124. *See also* Gradualism
  molecular biology and, 67–69
  natural selection in, 51–52, 53,
  54–58. *See also* Natural selection
  subtheories of, 52–58
Darwin's Dangerous Idea (Dennett),
  2, 88–92, 113, 136–137
  Gould's critique of, 93–97
Dating
  carbon, 21

  of Earth, 16–17
  of fossils, 21–22
  radioactive, 18–21
Dawkins, Richard, 1, 2, 75–78, 81,
  92, 141, 142, 185–188, 192–193,
  204, 227, 230
  views of: on natural selection,
  5–6, 75–76, 78–87, 92, 93;
  on punctuated equilibrium,
  111–113, 211–212; on species
  sorting, 113, 212
DDT, resistance to, 39–40
Dennett, Daniel, 1–2, 88–92, 93, 137,
  188, 190, 193, 227
  views of: on natural selection,
  88–97; on punctuated equilibrium, 113–115, 116–118,
  211–212; on species sorting, 115,
  212
Devonian period, 32
Dickermann, Mildred, 164–165
Dinosaurs, 33–34
  extinction of, 25, 34
DNA, 26–27
  chloroplast, 27, 28
  mitochondrial, 27, 28
  replication of, 68
  similarity of, demonstration of,
  70–71
  structure of, discovery of, 63–65
DNA-DNA hybridization, 70–71
Dobzhansky, Theodosius, 226
Double helix, discovery of, 63–65

Earth
  asteroid collisions with, 25, 34,
  67–68, 80–81, 223
  dating of, 16–17

formation of, 24–25
Ecological isolation, speciation and,
218–219, 223–224, 233–235
Edge of chaos, 135
Ediacarian fossils, 30
Einon, Dorothy, 164
Eldredge, Niles, 2, 79, 80, 96,
104–105, 106, 227
as evolutionary pluralist, 78–87,
92–98, 118–119, 141–142, 155,
206, 230–233
punctuated equilibrium theory of,
79–82, 106–118, 123–124, 183,
209–212. See also Punctuated
equilibrium
species sorting theory of, 5,
107–111, 113, 123–124, 204,
211, 212, 224
views of: on evolutionary psychology, 185–187, 203–204; on
natural selection, 4–6, 132; on
sociobiology, 185–187
Electrophoresis, 64–65
Emergent properties, of complex
systems, 127, 130–131, 203–206
Environmental adaptation, natural
selection and, 55–57
Eukaryotic cells, 26–27
evolution of, 27–29
Eukaryotic fossils, 27–28
Evolution
of bacteria, 25–26, 31
in Cambrian period, 30–31
computer simulations of, 141–153,
205
contingency in, 119–122
of eukaryotic cells, 27–29
extraterrestrial, 142–143

as fact, 45
fossil record of, 13–45. See also
Fossil(s); Fossil record
gene duplication in, 151
homology and, 44
of humans, 34–37
of mammals, 32–34
of marine animals, 30–32
microevolution and, 37–40
of multicellular animals, 29–31
of new species. See Speciation
ongoing research in, 222–226
parallel, 212–216, 234
of plants, 27, 28, 31
prey-predator interactions in,
146–147
of reptiles, 32–33
self-organization in, 136–137
Evolutionary biology
computer models in, 8–9. See also
Computer models
controversy in, 1–11, 235–237
ongoing research in, 222–226
recent advances in, 9–10
research directions in, 235–237
Evolutionary branching, 34–35,
69–73
dating of, 71–72
human-chimpanzee, 71–72
Evolutionary change, rate of, 9–10,
213–216, 219–222, 233–234. See
also Punctuated equilibrium
Evolutionary hierarchy, as miscon-
ception, 32
Evolutionary pluralism
vs. reductionism, 1–2, 5–6, 78–87,
92–98, 118–119, 185–193,
203–204, 230–233

Evolutionary pluralism *(continued)*
 sciences of complexity and,
 154–155
Evolutionary progress, 77
Evolutionary psychology, 6–7,
 157–193, 235–236
 adaptionism and, 84–85, 183–185
 controversy over, 179–193
 cooperative behavior and, 171–175
 vs. cultural relativism, 167–168
 Eldredge's critique of, 185–187
 future directions in, 202–203
 genetics and, 201–202
 Gould's critique of, 179–184,
 187–193
 language and, 183–184
 mating/sexual behavior and,
 162–169
 mental modules and, 6, 169–171,
 174–175, 180–181, 203, 208–209
 neural circuitry theory and,
 168–171, 174–175, 180–181,
 200–201
 principles of, 170–171
 rape and, 178–179
 social exchange and, 175–179
 sociobiology and, 180–181
 Wason selection task and, 175–178
Evolutionary stasis, 103–105, 114,
 115–116, 211
 habitat tracking and, 103, 116
Evolutionary theory. *See* Darwinian
 theory
Evolvability, 151–152, 231–232
Exaptations, 84
Exoskeletons, 31
The Extended Phenotype (Dawkins),
 87

Extinction
 contingency in, 119–120
 of dinosaurs, 25–26, 80
 mass, 25–26, 80–81, 120
 oxygen levels and, 29
 Permian, 80–81, 120
 punctuated equilibrium and,
 79–82
 in species sorting, 124
Extraterrestrial evolution, 142–143
Eye, evolution of, 56–57, 58

Fisher, Ronald, 61, 225
Fossil(s), 13–45
 bacterial, 28
 Cambrian, 30–31
 creation of, 22–24
 dating of, 21–22
 of earliest organisms, 25–26
 early theories of, 14–18
 Ediacarian, 30
 eukaryotic, 27–28
 gradualism and, 99–103
 interpretation of, 14–18
 Ordocivian, 31–32
Fossil record, 226–227
 gaps in, 99–103
 stasis in, 103–105
Freeman, Derek, 165–166

Gates, Bill, 15
Gene(s)
 interaction of, 128–137
 as replicators, 87–88
 selfish, 75–78
Gene duplication, 151
Genetic distance, interspecies, 71
Genetic drift, 212–216, 217, 234

Genetic mutations
  beneficial, 220–222
  effects of, 10, 220–222, 235
  natural selection and, 58–63, 108,
    219–222
Genetic networks, 128–137
  computer simulations of, 206
Genetics
  vs. gradualism, 60–63
  mathematical, 216–217, 224,
    225–227
  natural selection and, 58–60,
    75–78, 87–88
Genetics and the Origin of Species
  (Dobzhansky), 226
Genome
  as complex system, 129, 132–137
  mapping of, 66–67, 236
Geographic isolation, speciation and,
  101–103, 108–109, 211, 215,
  217, 219, 233–235
Geologic time scale, creation of, 17
Gorillas, 71
Gould, Stephen Jay, 1–2, 49–50, 106,
  227, 235
  as evolutionary pluralist, 1–2, 5–6,
    78–87, 92–98, 118–119,
    141–142, 155, 206, 230–233
  punctuated equilibrium theory of,
    79–82, 106–118, 183, 209–212.
    See also Punctuated equilibrium
  spandrel theory of, 6, 7, 83–84,
    122–123, 182–183, 206–209,
    224, 229–230
  species sorting theory of, 5,
    107–111, 113, 123–124, 204,
    211, 212, 224
  views of: on contingency, 119–122;

Dawkins' critique of, 78–87;
  Dennett's critique of, 88–92; on
  evolutionary psychology, 6–7,
  179–184, 187–193, 201,
  203–204; on mental modules,
  208–209; on natural selection,
  4–6, 78–97, 132; on sociobiology,
  179–181, 187–193
Gradualism, 51, 53, 99–124
  evolutionary stasis and, 103–105
  fossil record and, 99–103
  vs. genetics, 60–63
  mutational effects and, 220–222,
    235
  vs. punctuated equilibrium, 79–82,
    106–118, 183. *See also*
    Punctuated equilibrium
  vs. saltationism, 51, 99
Group behavior, complexity of,
  125–126
Group selection, 77
Guanine, in protein synthesis, 63–64

Habitat tracking, 103, 116
Haldane, J.B.S., 61, 225
Halstead, Lionel Beverly, 110
Harris, Henry, 65
Heredity, 58–60. *See also* Genetics
  natural selection and, 58–63
Hierarchy, evolutionary, as miscon-
  ception, 32
Hippocampus, 200
Holland, John, 135
Hominids, 34
Homo erectus, 36
Homo habilis, 34, 35–36
Homologies, 44, 73
Homo neanderthalis, 36–37

Homo sapiens, 36
Hooker, Joseph, 49, 52
How the Mind Works (Pinker),
182–184
Hubby, J.L., 64
Huey, Raymond, 213–214
Humans, evolution of, 34–37
Huxley, T.H., 62
as Darwin's champion, 2, 48–49,
93
debate with Wilberforce, 47–50
Hybridization, 42–43

"In Gratuitous Battle" (Gould), 187
Inheritance. *See also* Genetics
natural selection and, 58–63
Insecticides, resistance to, 39–40
Intelligence. *See also* Brain; Neural
circuitry
evolution of, 170–171, 201–202
spandrels and, 201
Internet resources, 245–251
Introns, 67

Johannsen, Wilhelm, 60
Jones, Corbin, 220–221, 222
Just-so stories, 84–85, 183, 184, 206

Kauffman, Stuart, 8–9, 129–142,
152–153, 154, 206, 231
views of: on gene interactions,
128–137; on origin of life,
137–141
Kettelwell, H.B.D., 39

Langton, Chris, 135
Language
cerebral processing of, 196–200

development of, 183–184
mental modules for, 198–200
as spandrel, 201
The Language Instinct (Pinker),
197, 202
Leclerc, Georges-Louis, 16–17
Leibniz, Gottfried, 85
Leicester Codex, 15
Lenski, Richard, 223
Lewontin, Richard, 5–6, 65, 83–87,
96, 122, 180, 182–183, 227
spandrel theory of, 6, 7, 83–84,
122–123, 182–183, 206–209,
229–230
Life, origins of, 25–27, 67–69
autocatalytic nature of, 137–141
computer simulations of, 206
extraterrestrial, 223
Lipids, 27–28
Logan, Graham A., 27
The Logic of Scientific Discovery
(Popper), 175–176

Mammals, evolution of, 32–34
Mapping, genome, 66–67, 236
*Margaret Mead and Samoa: The
Making and Unmaking of an
Anthropological Myth*
(Freeman), 165–166
Marine animals, evolution of, 30–32
Mars, origin of life on, 223
Mathematical genetics, 216–217,
224, 225–227
Mating/sexual behavior
animal, 157–160
cultural influences on, 160–162,
165–167
gender differences in, 162–164

genetic basis of, 162–164,
166–167, 171
human, 160–169
natural selection and, 168–169
parental investment and, 158
rape and, 178–179
Maynard Smith, John, 1, 2, 136, 141,
142, 191, 204, 227
views of: on evolutionary stasis,
110–111; on natural selection,
5–6, 92, 94, 96, 110; on punctuated equilibrium, 110–111
Mayr, Ernst, 52, 53, 67, 184
Mead, Margaret, 160–162, 165–167
Memory, 200
Mendel, Gregor, 59–60
Mental modules, 6, 168–171,
174–175, 180–181, 203,
208–209. *See also* Neural
circuitry
controversy over, 200–203,
208–209
for language, 198–200
Microevolution, 37–40
Migration, speciation and, 101–103,
108–109, 211, 215, 217, 219,
233–235
Mitochondria, 27, 28
Molecular biology, 63–69
Darwinian theory and, 67–69
principles of, 63–67
The Moral Animal (Wright), 190
Morganucodontids, 32–33
Mullis, Kary, 66
Multicellular animals, evolution of,
29–31
Mutations
beneficial, 220–222

effects of, 10, 220–222, 235
natural selection and, 58–63, 108,
219–222

Nagel, Laura, 215
A Natural History of Rape (Palmer),
178–179, 181–182
"Natural Language and Natural
Selection" (Pinker & Bloom),
182–184
Natural selection, 51–52, 53, 54–58
as algorithmic process, 89–92
brain function and, 170–171
controversies about, 3–6, 78–82
Dawkins' views on, 5–6, 75–76,
77, 78, 81, 82, 83
Dennett's views on, 88–97
edge of chaos and, 135
environmental adaptation and,
55–57
evolutionary psychology and,
170–171, 180–181
genetics and, 58–63, 136
Gould's views on, 4–6, 78–97
individual benefits of, 172
mating behavior and, 168–169
mutations and, 58–63, 108,
219–222
principles of, 3–4
selfish gene theory and, 75–78
as sole vs. main evolutionary force,
73–74, 76–97, 132, 136–137,
204, 216, 234–235
species sorting and, 5, 107–111,
123–124, 204, 211, 212, 224
without adaptation, 86
Neanderthals, 36–37
Neo-Darwinian synthesis, 62–63, 112

Network Tierra, 149–153
Neural circuitry
    controversy over, 200–203,
        208–209
    evolutionary psychology and,
        168–171, 174–175, 180–181
    for language, 198–200
    for social exchange, 171–179
Neurobiology, 168–171, 174–175,
    180–181, 200–203
Newton, Isaac, 132
The New York Review of Books, 92,
    95, 96, 116, 117, 188, 190
Nucleotides, in protein synthesis,
    63–64

On the Origin of Species (Darwin), 3,
    46, 47, 51, 52, 73, 92, 104, 225
Ordocivian period, 31–32
Organelles, 27–29
Organic chemicals
    autocatalysis among, 137–141
    sources of, 68–69
Orr, H. Allen, 2, 220, 221, 222
Owen, Richard, 47, 48
Oxygen levels, evolution and, 29–31

Paleontology. *See* Fossil(s); Fossil
    record
Palmer, Craig T., 6, 178–179,
    180–181
Panglossian paradigm, 84–85, 183,
    184, 206
Parental investment, 158
The Pattern of Evolution (Eldredge),
    115
Patton, James, 217
PCR (polymerase chain reaction), 66
Pelycosaurs, 33

Permian extinction, 80–81, 120
Pesticides, resistance to, 39–40
PET scans, 199
Photosynthesis, 29
Physical attractiveness, mating
    behavior and, 168–169
Pinker, Steven, 2, 182–184, 190, 197,
    198, 202, 208, 211
Plant cells, 27
Plants
    chloroplasts of, 27, 28
    hybridization of, 42–43
    migration of, 102–103
    terrestrial, evolution of, 31
Pluralism
    vs. reductionism, 1–2, 5–6, 78–87,
        92–98, 118–119, 185–193,
        203–204, 230–233
    sciences of complexity and,
        154–155
Polymerase chain reaction (PCR), 66
Popper, Karl, 175–176
Primates. *See also* Apes
    evolution of, 34–37
Primordial life, 25–27
Proteins
    electrophoretic studies of, 65–66
    research directions for, 236
    self-replication of, 68, 69
    synthesis of, 63–64; primordial,
        68–69
Psychology, evolutionary. *See*
    Evolutionary psychology
Punctuated equilibrium, 79–82,
    106–118, 123–124, 209–212,
    219–220
    complexity sciences and, 141–142
    in computer simulations, 205
    critiques of, 110–115, 183

defense of, 115–118
evidence for, 212–216
evolutionary stasis and, 103–105,
  114, 115–116, 211
genetic drift and, 212–216
speciation and, 79–82, 106–118,
  123–124, 183, 211, 215–216,
  223–224

Radioactive dating, 18–21
Radioactive decay, 19
Ramapithecus, 72
Rape, 178–179
Ray, Thomas, 8–9, 154, 205, 232
  Tierra experiment of, 143–153,
  205
Reciprocity, 173–174
Reductionism, 1–2, 5n, 82–83, 141,
  142, 202–204, 227, 230–233
  vs. evolutionary pluralism, 1 2,
  5–6, 78–87, 92–98, 118–119,
  185–193, 203–204, 230–233
  mathematical genetics and,
  216–217
Reinventing Darwin (Eldredge),
  107
Reproduction. *See also* Mating/sexual
  behavior
  natural selection and, 54–55
Reproductive isolation, 40–42
Reptiles, evolution of, 32–33
RNA, 64
  self-replication of, 68–69
Rundle, Howard, 215

Saltationism, 51, 99, 116, 117
Sarich, Vincent, 71
Schemske, Doug, 221, 222
Schluter, Dolph, 214–215, 218–219

Sciences of complexity. *See*
  Complexity sciences
Scientific controversy
  benefits of, 203, 227–230, 237
  evolution of, 228–229
  resolution of, 118–119
  value of, 11
Scientific theory, definition of, 45
Sedimentary rock
  formation of, 15
  fossils in, 15–18, 22–24
  geological epochs and, 16–18
  strata of, 15–16
The Selfish Gene (Dawkins), 1, 75
Self-organization, in evolution,
  136–137
*Sex and Temperament in Three
  Primitive Societies* (Mead), 161
Sexual behavior. *See* Mating/sexual
  behavior
Seysenberg, Erich Tschermark von, 60
Sheldon, Peter, 110
Sivapithecus, 72
Sleep, Norman, 223
Social exchange, neural circuitry for,
  171–179
Sociobiology, 179–181, 185
Spandrels, 6, 7, 83–84, 122–123,
  182–183, 201, 206–209, 224,
  229–230
  in brain, 201, 208 209
Speciation, 10, 40–45, 211, 215–216
  ecological isolation and, 218–219,
  223–224, 233–235
  geographic isolation and,
  101–103, 108–109, 211, 215,
  217, 219, 233–235
  as gradual process, 100–101. *See
  also* Gradualism

Speciation, *(continued)*
  hybridization and, 42–43
  natural selection and, 107–111
  punctuated equilibrium and,
    79–82, 106–118, 123–124, 183,
    211, 215–216, 223–224
  reproductive isolation and, 40–42
Species
  extinction of. *See* Extinction
  genetic distance between, 71
  multiplication of, 52, 53
  new. *See* Speciation
Species correspondence, 43–44
Species homology, 44
Species sorting, 5, 107–111, 123–124,
    204, 211, 212, 224
  Dawkins' critique of, 113
Spencer, Herbert, 52
Standard Social Science Model, 167
Steno, Nicholas, 16
Summons, Roger E., 27
Survival of the fittest, 52

Tetrapods, 43–44
Theory, scientific, definition of, 45
Theraspids, 33
The Third Culture (Brockman), 87
Thornhill, Randy, 6, 178–179,
    181–182
Thymine, in protein synthesis, 63–64
Tierra simulations, 144–151, 153,
    205
Tiselius, Arne, 64

Tooby, John, 2, 167–170, 176–178,
    184–185, 190–192, 200, 201,
    203, 208
Tragopogan, new species of, 41–42
Trilobites, 103–105
Triplets, nucleotide, 64
Trivers, Robert, 158–159
Tschermark von Seysenberg, Erich, 60
Turner, J.R.G., 110

Ultra-Darwinists, 227

Vinci, Leonardo da, 14–15
Virtual computers, 144
Voltaire, François, 84–85
Vrba, Elizabeth, 84
Vries, Hugo de, 60

Wason, Peter, 175
Wason selection task, 175–178
Watson, James, 63
Web sites, 245–251
Wernicke's aphasia, 197–198
Wernicke's area, 197, 199
Wilberforce, Samuel, 47–50
Williams, George C., 76–77, 78, 87,
    96n
Wilson, Allan, 71
Wilson, Edward O., 179–180,
    185–188, 193
Wright, Robert, 190
Wright, Sewall, 61, 225